Port
Essington

Darwin

Torres Strait

Cape York

Gulf of
Carpentaria

Victoria R

BARKLY TABLELAND

Cape
Keer Weer

GREAT DIVIDING RANGE

Endeavour R

Cairns

NORTHERN TERRITORY

ATHERTON
TABLELAND

Townsville

GREAT DIVIDING RANGE

Mackay

QUEENSLAND

Nogoa R

Dawson R

Bustard Bay

Burnett R

Wide Bay

Mary R

SOUTH AUSTRALIA

Condamine R

Darling
Downs

Brisbane, & Moreton Bay

FLINDERS RANGES

Richmond R

Namoi R

Clarence R

Darling R

Liverpool
Plains

Macleay R
Port Macquarie

Eyre Peninsula

Port
Augusta

NEW SOUTH WALES

Lord •
Howe I

Spencer Gulf

Lachlan R

Maitland

Blackheath

Newcastle, & *Hunter R*

Gulf of St Vincent, & Adelaide

Murrumbidgee R

Sydney, & Port Jackson
Botany Bay

Argyle County

Kangaroo I

VICTORIA

Murray R

GREAT
DIVIDING RANGE

Kow
Swamp

GRAMPIANS

Melbourne

Port Fairy
Warrnambool

King I

Bass Strait

Preservation I

Furneaux Is
Cape Barren

TASMANIA

Derwent R

Adventure Bay
Hobart, Blackmans Bay, & Storm Bay

EUROPEAN EXPLORERS were amazed by the strange animals they found in Australia; their accounts make fascinating reading; their sketches are charming, if often inaccurate. Here is the first collection of this material, species by species, gleaned from books and other sources that date back to the 1600s. It includes material on more than 60 species, many with the first recorded European sighting, many with the first scientific description, and all with up-to-date biological information. The bibliography lists references to almost every voyage of discovery or exploration into Australia during the eighteenth and early nineteenth centuries.

PETER STANBURY is Director of The Macleay Museum, The University of Sydney, and author or editor of a number of books, including *Looking at Mammals, One Hundred Years of Australian Scientific Explorations, Ten Thousand Years of Sydney Life,* a history of Australian photography, and a number of scientific papers. He is editor of the Australian Museum's Association journal, *Kalori.*

GRAEME PHIPPS is the ornithologist at The Macleay Museum, President of the Avicultural Society of New South Wales, editor of *The Avicultural Review,* and a member of the Royal Zoological Society of New South Wales and other natural history societies.

AUSTRALIA'S ANIMALS DISCOVERED

AUSTRALIA'S ANIMALS DISCOVERED

Peter Stanbury and Graeme Phipps
The Macleay Museum, The University of Sydney

PERGAMON PRESS
SYDNEY · OXFORD · NEW YORK · TORONTO · PARIS · FRANKFURT

First published 1980

Copyright © 1980 Peter Stanbury and Graeme Phipps

Pergamon Press (Australia) Pty Ltd
19a Boundary Street, Rushcutters Bay, N.S.W. 2011, Australia.
Pergamon Press Ltd
Headington Hill Hall, Oxford OX3 0BW, England.
Pergamon Press Inc.
Maxwell House, Fairview Park, Elmsford, N.Y. 10523, U.S.A.
Pergamon of Canada Ltd
Suite 104, 150 Consumers Road, Willowsdale, Ontario M2J 1P9, Canada.
Pergamon Press GmbH
6242 Kronberg-Taunus, Pferdstrasse 1, Federal Republic of Germany.
Pergamon Press SARL
24 rue des Ecoles, 75240 Paris, Cedex 05, France.

Designed by Robin James
Typeset in Australia by S.A. Typecentre Pty Ltd
Printed in Singapore by Toppan Printing Co.(S) Pte Ltd

National Library of Australia Cataloguing in Publication Data:
    Stanbury, Peter J.
    Australia's animals discovered.
    Index
    Bibliography
    ISBN 0 08 024796 2
    1. Zoology — Australia. 2. Animals. I. Phipps, Graeme,
    joint author. II. Title.
    591.994

Contents

AUSTRALIAN MUSEUM

This is to Certify that the Bearer Mr. George Masters Assistant Curator of the Australian Museum, has been commissioned by the Trustees thereof, to proceed to Western Australia, for the purpose of collecting Specimens of Natural History.

Dated at Sydney, New South Wales. this 5th day of September 1868.

E. Deas Thomson.
Crown Trustee and Chairman of the Board.

Gerard Krefft
Curator and Secretary
Signature of the Bearer
George Masters. Assistant Curator

An engraved letter of introduction from the Trustees of the Australian Museum issued to an authorised collector in 1868. The collector, George Masters, was later to become the first curator of The Macleay Museum.

Preface

For more than 40 000 years the Australian Aboriginals have observed the ways of the continent's animals. Their legends of the Dreamtime often told the story of how animals came into being or acquired certain characteristics. The legends were passed from generation to generation, and the oral tradition reinforced by paintings or rock engravings.

European philosophers of the Middle Ages were quite familiar with the concept of a spherical earth, which included for many a mythical *Terra Australis Incognita* – an unknown land to the south. One of the earliest maps to show any close resemblance to the Australian coastline was the French Dieppe map, an ornamental map drawn for King Henry VIII between 1530 and 1550; it is thought that its origin was the pilots' charts of Portuguese mariners.

The northern and western coasts of Australia were mapped by the Dutch between 1616 and 1627, and called 'New Holland' — with the unknown part beyond New Guinea being known as 'Terra Australis'. Abel Tasman in 1642 discovered Tasmania, and named it Van Diemen's Land, after the Governor-General of the Dutch East Indies. It retained this name officially until 1855 when it was renamed 'Tasmania', which had been in common usage since 1823.

In 1768 Captain James Cook was sent in the *Endeavour* to observe the transit of Venus from Tahiti, and to enquire into the southern continent. He mapped the east coast and named it 'New South Wales'. When the commission was given to Captain Arthur Phillip on 2 April 1787 as 'Captain-General and Governor-in-Chief of New South Wales', his area of jurisdiction extended from Cape York in the north to the southern extremity of the continent, and inland to 135°E. All land to the west was still named New Holland, which in official circles was sometimes taken to be this western part, and at other times to be the entire continent, but it faded from use in the 1820s in favour of 'Australia'.

The first European accounts of Australian animals and flora were by sailors; then by frequently ill-equipped and under-nourished explorers, by settlers preoccupied with tending a crop, and by convicts on the run. They had no time to study the habits of the animals they saw and often came to the wrong conclusions. Later, when scientific studies had been completed in museums and as more reports came in, the names and relationships of animals were then changed. It is because of this that names in early reports do not always agree with those in use today.

Research for a book like this starts simply. To trace the first sighting, the first description and the first illustration, is a matter of tracking down references in books. For example, the current scientific name of the frilled lizard is *Chlamydosaurus kingi* Gray. This tells us that the animal was first described scientifically by someone called Gray; but where? One place worth looking was in the *Catalogue of Lizards in the British Museum (Natural History)* compiled by G. A. Boulenger, and published in three volumes between 1893 and 1896. These technical works mentioned that the description was by *Gray, in King's Voy. Austr. ii.*

p. 424. Looking through all the 'King's' in a library catalogue, we found a book by Phillip Parker King: *Narrative of a Survey of the Intertropical and Western Coasts of Australia Between the Years 1818 and 1822 . . .* 'With an Appendix containing Various Subjects Relating to Hydrography and Natural History'. This two-volume work was published in 1827. The known range of the frilled lizard *is* in the area of the title of this book; and the 'ii' in the *Catalogue of Lizards* reference suggests the second volume of a series. The book was obtained from the Rare Book Library in the Fisher Library at the University of Sydney, and contains an appendix written by John Edward Gray of the British Museum, in which he described the new animal.

It was usual for explorers to send all material of a scientific nature to institutions such as the British Museum of Natural History for classification and description; but in many cases these descriptions are entirely in Latin. This time it was not, luckily, and a clue was given in the scientific description about who supplied the new species. The first part of the work is a 'narrative' or 'journal'; and on looking up the pages for the period the expedition was in the locality where the collection was made, it was found that on 8 October 1820 at Careening Bay, north-western Australia, King recorded that 'Mr. Cunningham found a very curious species of lizard'.

Allan Cunningham was a botanist on the expedition, collecting plants for 'His Majesty's Botanical Gardens at Kew', and his journal report – also fortunately in the same book – contains remarks about the first sighting.

The frilled lizard is a simple example. Much detective work is needed in some circumstances. But it is enjoyable doing such research from primary sources, especially when a pattern begins to emerge.

It is also while researching such sources that details about the personalities of the people emerge. What, for example, was a man like William Dampier doing in Australian waters? Many people do not know that he was a pirate and his reasons for being in the East Indies originally were unlikely to be to admire the natural attributes of the area. He wrote a book about his travels, *New Voyage Round the World* published in 1697; the book prompted the British Admiralty to hire him, an ex-buccaneer with no leadership experience, to conduct an expedition of discovery to the vast area south and south-east of the Dutch East Indies, of which nothing was then known. They purchased a rotten ship, the *Roebuck*, and equipped the expedition as cheaply as possible, with an inferior crew and without officers of quality to make up for Dampier's deficiencies. The episodes of the expedition make fascinating reading.

And what were many well-educated people doing in what was then the end of the world? Whatever their reasons, we can only be glad that such characters came and recorded the details of their travels.

Sir Joseph Banks has been called by some people 'the father of Australia'. This very wealthy man accompanied Cook on the *Endeavour* expedition. One must admire Cook for keeping such a diverse array of personalities together and working in reasonable harmony on board ship. To Banks, matters scientific had complete precedent over any other consideration (apart from his own comfort): for the second expedition of Cook's, the collier *Resolution* was converted by the addition of a superstructure giving an extra deck and a great 'cabbin' so that Banks and his entourage could work in style. These alterations cost between £6 000 and £14 000, depending on whose reports are to be believed. But the superstructure made the ship so unseaworthy that it had to be removed; the letters between Banks and the Admiralty concerning the matter show that Banks eventually declined to go on the expedition. Yet he retained a great interest in the continent, and it was he who suggested that Botany Bay be used as the site for a penal colony.

In 1786 Lord Sydney authorised the despatch of an expedition to form a settlement in Botany Bay, and the First Fleet of eleven ships carrying convicts, officials and marines sailed in 1787, finally settling in the adjacent harbour of Port Jackson on 26 January 1788.

From the early days of the colony almost until his death, Joseph Banks was sent plants, parts of plants, and animals – almost anything of a botanical or zoological nature was deemed to be of interest to him. He also arranged for collectors to go to New Holland, and some very able people were sent. (He also sent people considered too much of a nuisance at home.) Looking back, it seems there was never a dull moment in early Sydney Town; one can almost feel the Governor wince at the despatches received, accompanying yet another person from the mother country.

The species chosen for inclusion in this book are all vertebrate animals: mammals, birds, fish, lizards and snakes. While there are invertebrates unique to Australia, we chose the larger, better-known animals because they trace the story of the early settlement and the development of Australian zoology. The spiny anteater, for example, was beautifully illustrated by William Bligh – better known as a great navigator and survivor of the mutiny on the *Bounty*, and as a Governor of the colony. Most of the raw material of this book was first published in Europe: John Gould and his wife began *The Birds of Australia* in England but decided they must visit Australia to complete their work. Upon their return to England, Gould finished one of the greatest expositions of the fauna of any country.

We hope that the readers of our book become absorbed, as we did, with the fascinating coincidences of Australian history, the colonial days, the explorers, the pioneer zoologists, and early discoveries of Australia's unique animals.

Peter Stanbury
Graeme Phipps
1980

ACKNOWLEDGEMENTS

Alan Davies, our photographer, proof-reader, bibliographer and constructive critic, helped so much with this book that without him it is doubtful it would have been completed. We are also grateful to our colleagues at the Macleay Museum, Lydia Bushell and Judy Leon, who helped mount the exhibition 'Australia's Animals: Who Discovered Them?', which gave us the idea for this book. Jack Mahoney, Department of Geology and Geophysics, University of Sydney, discussed the literature of Australian mammals with us and gave us much moral support.

The following libraries, institutions and individuals allowed us to look at their collections and use selected items. We thankfully record their assistance: The Australian Museum; Associate Professor D. F. Branagan; The British Museum (Natural History); Mr J. K. Clegg; William Collins Pty Ltd; Fisher Library, University of Sydney; Mitchell Library, State Library of New South Wales; The Museum of Applied Arts and Sciences; National Library, Canberra; and Mr C. J. Percival.

This stylised illustration appeared in *A History of the Earth and Animated Nature* by the poet Oliver Goldsmith.[41] The edition we viewed was published around 1840, but the first edition was produced in 1774. Note the triplets in the pouch of one of the kangaroos. Kangaroos have four teats, so it is possible, though extremely rare, to have three young. The animals pictured were captioned: '1. Lord Derby's Kangaroo. 2. Aröe Kangaroo. 3. Parry's Kangaroo. 4. Woolly Kangaroo. 5. Brush Tailed Kangaroo. 6. Rat-tailed Hypsiprymnus. 7. Rabbit-eared Perameles [this is, in fact, a rabbit-eared bandicoot].'

The word 'kangaroo' is now used only for the largest macropods. Smaller macropods are called wallaroos, then wallabies, potoroos, and finally rat-kangaroos.

Mammals

The astonishment that early European settlers and naturalists must have experienced when discovering what amounted to a 'lost world' of animals can only be imagined today. When the first explorers sent home specimens of the mammals, it was thought at first that some kind of taxidermist's practical joke was being played, until it was realised that the southern continent was an attic-full of mammals superseded in other countries. And not merely superseded – for these primitive mammals had continued to evolve to fit the environment.

The mammals of the world are divided into three major groups, and Australia and the nearby islands are the only places in the world where these three groups occur together. Geographically isolated, the peculiarly Australian monotremes (the platypus and the spiny anteater) and marsupials (kangaroos, possums and others) developed to look different from their ancestors.

The monotremes are the only living representatives of a primitive group of reptile–like mammals long extinct elsewhere. They show many reptilian features: they lay eggs; their digestive and reproductive tubes open into a common chamber (monotreme means 'one opening'); their front limbs are held out in reptilian fashion from the body, rather than under it (as in a dog, for example); and the control of body temperature is poor compared to that of a dog or a human being.

The marsupials represent another evolutionary progression towards the complexity of sophisticated mammals. The young are born alive but in such an embryonic condition that they must continue their development in the pouch. Although this reproductive system might appear slow and ponderous, a device called 'delayed implantation' – by which fertilised eggs can be held in a state of arrested development, awaiting the vacation of the pouch by a previous baby, or cessation of drought conditions – means that marsupial reproduction has remarkable compensations for its apparent inefficiency.

The marsupials regarded as nearest the ancestral stock are the carnivorous ones, including the thylacine, the Tasmanian devil and the native cats. Thylacines were hunted into extinction by Europeans in Tasmania; and possibly by the Aboriginals and their dingoes on the mainland. The Tasmanian devil, however, is still common in Tasmania, and seems to be one of the few marsupials to contend successfully with the activities of the European. It is equivalent ecologically to the hyaena of Africa and Asia – both animals are scavengers with jaws for crushing bone.

Bandicoots belong to a group intermediate between the carnivorous marsupials and the herbivorous kangaroo-wombat-possum group. Wombats have traditionally been grouped with the possums, but it is probable that they are more closely related to the koala. Kangaroos, wallabies, potoroos, bettongs and other similar animals are grouped in the family *Macropodidae*. The largest members of the family, the red and grey kangaroos, may grow to a height of 2 metres; the smallest rat kangaroos are less than 250 mm high. More than fifty species of macropod live in Australia and New Guinea.

The dugong, dingo and flying fox and the native rats are examples of the third major group of mammals, the eutherians – the mammals that evolved and displaced the monotremes and marsupials in almost all parts of the world except Australia. The more primitive monotremes and marsupials survived because the Australian continent became isolated by sea early in mammalian evolution. More than fifty per cent of the Australian mammal fauna is in fact eutherian, but the Australian representatives are immigrants; either they arrived by their own efforts or they were brought to the continent. The dugong swam, flying foxes (bats) flew, and the Aboriginals brought the dingo some 10 000 years ago. No-one knows when the first rodents arrived, possibly on log rafts, but it was geologically quite late. These early rodents radiated, to give the vast array of native rats and mice ranging from the desert-dwelling *Notomys* to the otter-like water rats.

The Aboriginals themselves are thought to have arrived more than 40 000 years ago and were the first people to see Australia's unique fauna. They saw not only what the first white settlers saw, but also the disappearance of the megafauna – wombat-like animals the size of a rhinoceros, kangaroos eight times the mass of today's wallabies, and a carnivorous lion-sized marsupial.

Platypus

Ornithorhynchus anatinus

The unique platypus is found only in Australia. Like the spiny anteater, it is a mammal but shows some characteristics of more ancient animals.

When the first skin reached Europe, it was thought to be a hoax; and although in the nineteenth century it was suspected to lay eggs, final proof was not obtained until 1884. The female lays two eggs in a nesting hutch and curls around them until they hatch. Neither the platypus nor the spiny anteater has nipples, so the milk exudes from many pores which the young suck or lick. The young remain in the nest for about four months; they are then weaned and feed on aquatic insects, frogs, small fish and earthworms.

The platypus breathes air but can stay under water for about five minutes. Its beak is soft, flexible and sensitive – quite unlike the stiff appearance of museum exhibits. The male has a poisonous spur on its hind foot; the venom can kill a rabbit in ninety seconds.

It is a shy and solitary creature, found only in the eastern mainland and Tasmania.

An early European illustration of a platypus was published in Thomas Bewick's *General History of Quadrupeds*[13] only three years after its discovery, under the title of 'an amphibious animal'.

First sighting and capture   **November 1797**

The Kangaroo, the Dog, the Opossum, the Flying Squirrel, the Kangaroo Rat, a spotted Rat, the common Rat, and the large Fox-bat (if entitled to a place in this society), made up the whole catalogue of animals that were known at this time, with the exception which must now be made of an amphibious animal, of the mole species, one of which had been lately found on the banks of a lake near the Hawkesbury. In size it was considerably larger than the land mole. The eyes were very small. The forelegs, which were shorter than the hind, were observed, at the feet, to be provided with four claws, and a membrane, or web, that spread considerably beyond them, while the feet of the hind legs were furnished, not only with this membrane or web, but with four long and sharp claws, that projected as much beyond the web, as the web projected beyond the claws of the fore feet. The tail of this animal was thick, short, and very fat; but the most extraordinary circumstance observed in its structure was, its having, instead of the mouth of an animal, the upper and lower mandibles of a duck. By these it was enabled to supply itself with food, like that bird, in muddy places, or on the banks of the lakes, in which its webbed feet enabled it to swim; while on shore its long and sharp claws were employed in burrowing; nature thus providing for it in its double or amphibious character. These little animals had been frequently noticed rising to the surface of the water, and blowing like the turtle.

Hawkesbury River, N.S.W.   **David Collins**[28]

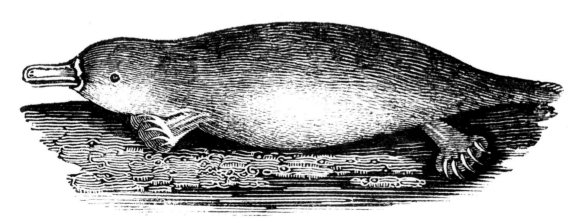

13

First published description   1799

. . . Of all the Mammalia yet known it seems the most extraordinary in its conformation, exhibiting the perfect resemblance of the beak of a Duck engrafted on the head of a quadruped. So accurate is the similitude, that, at first view, it naturally excites the idea of some deceptive preparation by artificial means: the very epidermis, proportion, serratures, manner of opening, and other particulars of the beak of a shoveler, or other broad-billed species of duck, presenting themselves to the view: nor is it without the most minute and rigid examination that we can persuade ourselves of its being the real beak or snout of a quadruped.

Near Sydney   **George Shaw**[97]

Not a hoax   1800

Of this most extraordinary genus the first description appeared in the *Naturalist's Miscellany*; but as the individual there described was the only one which had been seen, it was impossible not to entertain some distant doubts as to the genuine nature of the animal, and to surmise, that, though in appearance perfectly natural, there might still have been practised some arts of deception in its structure. I, therefore, hesitated as to admitting it into the present History of Quadrupeds. Two more specimens, however, having been very lately sent over from New Holland, by Governor Hunter, to Sir Joseph Banks, the suspicions before mentioned are now completely dissipated.

Near Sydney   **George Shaw**[100]

This pair of platypuses from the *Penny Cyclopaedia* of 1840[87] exhibits a certain amount of artistic licence; the curled positions of this pair would be more appropriate in a nesting burrow.

First popular account   1800

[The platypus] is found in the fresh water lakes, which is about the size of a small Cat; it chiefly frequents the banks of the lakes; its bill is very similar to that of a Duck, and it probably feeds in muddy places in the same way; its eyes are very small; it has four short legs; the fore legs are shorter than those of the hind, and their webs spread considerably beyond the claws, which enables it to swim with great ease; the hind legs are also webbed, and the claws are long and sharp. They are frequently seen on the surface of the water, where they blow like a turtle: their tail is thick, short and very fat.

The natives say they sometimes see them of a very large size.

Near Sydney   **Thomas Bewick**[13]

First proof of egg-laying   24 August 1884

. . . I shot an *Ornithorhynchus* whose first egg had been laid; her second egg was in a partially dilated [uterus]. This egg, of similar appearance to, though slightly larger than, that of *Echidna*, was at a stage equal to a 36-hour chick. On the 29th [of August] I sent in the telegram 'Monotremes oviparous, ovum meroblastic' to the British Association at Montreal.

Burnett River, Qld   **W. H. Caldwell**[24]

Spiny anteater

Tachyglossus aculeatus

The spiny anteater is a mammal, grouped with the platypus; both animals lay eggs, suckle their young and are warm blooded, but their temperature ranges are much wider than most mammals. Zoologists consider they show the characteristics of both mammals and reptiles and are living representatives of the ancestors of some of the first mammals to evolve.

One common name for the spiny anteater, 'echidna', alludes to the hollow spur on the inner side of the lower hind limb in males; this organ is retractable and can inject a weak poison (the Latin *echidna* means adder or viper – that is, a poisonous being). However, the name was earlier given to some eels, and for consistency 'spiny anteater' is preferred as a common name.

The scientific name, *Tachyglossus*, means quick-tongued, referring to the 300-mm-long sticky tongue on which it catches termites and other ants. The mouth is at the tip of the hard beak.

The female lays an egg which is placed in her pouch. On hatching, the young has soft spines (which are modified hairs), and remains in the pouch for three weeks; by then the spines are hard. The mother hides the baby in a hollow, returning periodically to suckle it.

The spiny anteater can roll into a ball to form a defensive sphere of spines; it may be found throughout the mainland and in Tasmania.

In 1792 George Tobin was third lieutenant on William Bligh's second voyage to procure breadfruit for the West Indies, and to explore Torres Strait. Tobin sketched the spiny anteater in his notebook, which is now in Sydney's Mitchell Library.[111]

A naturalist's report   **9 February 1792**
The only animals seen, were the Kangoroo, and a kind of sloth about the size of a roasting pig with a proboscis two or three inches in length – On the back were short quills like those of the porcupine – This animal was roasted and found of a delicate flavour – The Kangoroos were so rapid in their motions they escaped all our guns.
Adventure Bay, Tas.   **George Tobin**[111]

William Bligh's skill as an artist is little known. But his excellent sketch of the spiny anteater was sent to Sir Joseph Banks, who arranged for this engraving to appear in the Royal Society Journal of 1802[57], to accompany the anatomical description of Professor Home.

An animal of a very odd form   10 February 1792
It was seventeen inches long, and has a small flat head connected so close to the shoulders that it can scarce be said to have a neck. It has no mouth like any other animal, but a kind of duck-bill, two inches long, which opens at the extremity and will not admit anything above the size of a pistol ball. It has four legs, and on each foot are very sharp claws; it has no tail but a rump not unlike a penguin's, on which are quills of rusty brown.
Adventure Bay, Tas.   William Bligh[57]

A scientific curiosity   July 1792
This extraordinary animal may well be considered amongst the most curious and interesting quadrupeds yet discovered; since it is not only an absolutely new and hitherto unknown species, but is also a most striking instance of that beautiful gradation, so frequently observed in the animal kingdom, by which creatures of one tribe or genus approach to those of a very different one. In its mode of life this animal beyond a doubt resembles the Myrmecophagae [anteaters], having been found in the midst of an ant-hill; for which reason it was named by its first discoverers the ant-eating porcupine . . . It will even burrow under a pretty strong pavement, removing the stones with its claws; or under the bottom of a wall. During these exertions its body is stretched or lengthened to an uncommon degree, and appears very different from the short or plump aspect which it bears in its undisturbed state.
New Holland   George Shaw[97]

Native cats

Dasyurus spp.

Tracks?   **1 May 1770**

We saw one quadruped about the size of a Rabbit . . . also the dung of a large animal that had fed on grass which much resembled that of a Stag; also the footsteps of an animal clawd like a dog or wolf and as large as the latter; and of a small animal whose feet were like those of a polecat or weesel. [Beaglehole suggests these may have been a bandicoot or kangaroo rat, kangaroo dung, dingo tracks, and native cat tracks.]

Botany Bay   **Joseph Banks**[8]

Named the 'quoll'   **August 1770**

Another was calld by the natives *Je-Quoll*: it is about the size and something like a polecat, of a light brown spotted with white on the back and white under the belly.

New South Wales   **Joseph Banks**[8]

Another cat   **March 1772**

The savages make use of bark from trees for cooking shell fish. There was little game, and we presumed that the fires made by the savages in this place had driven them inland. Our hunters met a tiger cat, and found several holes like those in a warren.

Blackman's Bay, Tas.   **Julian Crozet**[29]

A more detailed description   **1789**

The species is about the size of a large polecat, and measures from the tip of the nose to the setting on of the tail eighteen inches; the tail itself being nearly the same length. The visage is pointed in shape, and the whole make of the animal does not ill resemble that of the *Fossane*. The general colour of the fur is black, marked all over with irregular blotches of white, the tail not excepted, which has an elegant appearance, and tapers gradually to a point.

The situation of the teeth and jaws is much the same as in the rest of the genus, as may be seen in the upper part of the plate.

Port Jackson   **Arthur Phillip**[90]

This rarely seen marsupial was drawn for the journal of the colony's first Governor, Arthur Phillip, which was published in 1789.[90] The tail is incorrectly drawn too thin.

There are four species of the marsupial 'native cat', and members of one or more species are found in most forested regions of the continent and Tasmania. The early descriptions indicate

Two colour varieties of the eastern native cat are well shown in this illustration from John Gould's *Mammals of Australia*, published from 1845 and completed in 1863.[44]

that native cats were commoner than they are now, as feral European cats compete for the same prey, and as thickets are cleared for farms.

The eastern native cat, or eastern quoll, *Dasyurus viverrinus* gives birth to ten to twenty-four babies, but as there are only six or eight teats, many die in each generation. The young, after becoming furred, ride on their mother's back; the spotted coat acts as camouflage in the dappled light of the forest.

Tasmanian devil

Sarcophilus harrisii

The Tasmanian devil was named after its black colour, ungainly appearance, imagined savage disposition, and nocturnal activities (though it is fond of basking in the sun).

It has strong teeth and jaws and can demolish its prey, bones, skin and all. It is the largest marsupial carnivore (assuming the thylacine to be extinct) and is still relatively common in Tasmania. Like the thylacine, it once roamed mainland Australia but became extinct there before the arrival of Europeans.

Tasmanian devils mate in the spring; the three or four young are born in a nest of leaves and grass, and remain with the mother for about six months.

First description   **21 April 1807**

These animals were very common on our first settling at Hobart Town, and were particularly destructive to poultry, &c. They, however, furnished the convicts with a fresh meal, and the taste was said to be not unlike veal. As the settlement increased, and the ground became cleared, they were driven from their haunts near the town to the deeper recesses of forests yet unexplored. They are, however, easily procured by setting a trap in the most unfrequented parts of the woods, baited with raw flesh, all kinds of which they eat indiscriminately and voraciously; they also, it is probable, prey on dead fish, blubber, &c. as their tracks are frequently found on the sands of the sea shore.

Van Diemen's Land   **G. P. Harris**[55]

The adult Tasmanian devil is some 750-1000 mm long. Gould's life-like illustration[44] also shows the long whiskers, which act as sensors in the undergrowth.

The retreat of the devil   1863

The Ursine Sarcophilus was one of the first of the native quadrupeds encountered by the early settlers in Van Diemen's Land, from whom its black colouring and unsightly appearance obtained for it the trivial names of Devil and Native Devil. It has now become so scarce in all the cultivated districts, that it is rarely, if ever, seen there in a state of nature; there are yet, however, large districts in Van Diemen's Land untrodden by man; and such localities, particularly the rocky gullies and vast forests on the western side of the island, afford it a secure retreat. During my visit to the continent of Australia I met with no evidence that the animal is to be found . . . consequently Tasmania alone must be regarded as its native habitat. In its disposition it is untameable and savage in the extreme, and is not only destructive to the smaller kangaroos and other native quadrupeds, but assails the sheep-folds and hen-roosts wherever an opportunity occurs for its entering upon its destructive errand.

Tasmania   **John Gould**[44]

Tasmanian devils on the mainland   1912

Among the sand-hills which fringe the coast between Warrnambool and Port Fairy, Victoria, there are many relics of aboriginal inhabitants. Ashes and blackened soil mark the site of their camp fires. In addition to the shells there are also numerous bones lying on the surface of the sand. The most interesting example in the collection is that of the right mandible of *Sarcophilus ursinus*, which from its appearance, is very recent. It would appear then that the Tasmanian Devil survived till a very late period in this part of Victoria, and that it was contemporaneous with the Australian aborigine.

Warrnambool, Vic.   **D. J. Mahoney**[79]

This illustration of the Tasmanian devil was published with G. P. Harris's first description in *Transactions of the Linnean Society of London*.[55]

Thylacine

Thylacinus cynocephalus

Harris's 1807 sketch.[55]

First description   **30 March 1805**

An animal of a truly singular and nouvel description was killed by dogs the 30th March on a hill immediately contiguous to the settlement at Yorkton Port Dalrymple; from the following minute description of which, by Lieutenant Governor Paterson, it must be considered of a species perfectly distinct from any of the animal creation hitherto known, and certainly the only powerful and terrific of the carnivorous and voracious tribe yet discovered on any part of New Holland or its adjacent islands.

'It is very evident this species is destructive, and lives entirely on animal food; as on dissection his stomach was found filled with a quantity of kangaroo, weighing 5 lbs. The weight of the whole animal 45 lbs. From its interior structure it must be a brute peculiarly quick of digestion . . . stripes across the back 20, on the tail 3; 2 of the stripes extend down each thigh; length of the hind leg from the heel to the thigh . . . the stripes black; the hair on the neck rather longer than that on the body; the hair on the ears of a light brown colour, on the inside rather long. The form of the animal is that of the hyaena, at the same time strongly reminding the observer of the appearance of a low wolf dog. The lips do not appear to conceal the tusks.'

Yorkton, Tas.   Lt. Gov. William Paterson[105a]

A naturalist's report   **21 April 1807**

Only two specimens have yet been taken. It inhabits amongst caverns and rocks in the deep and almost impenetrable glens in the neighbourhood of the highest mountainous parts of Van Diemen's Land, where it probably preys on the brush Kangaroo, and various small animals that abound in those places. That from which this description and the drawing accompanying it were taken, was caught in a trap baited with kangaroo flesh.

Van Diemen's Land   G. P. Harris[55]

Early account   **January 1829**

Considerable numbers of the native hyena prowl from the mountains in search of prey among the flocks at night. The shepherd is therefore obliged, during the lambing season, either to watch his flocks during the night, or to enclose them in a fold . . . One of them measured six feet from the snout to the tail. The skin is beautifully striped with black and white on the back, while the belly and sides are of a grey colour. Its mouth resembles that of a wolf, with large jaws opening almost to the ears. Its legs are short in proportion to the body, and it has a sluggish appearance, but in running it bounds like a kangaroo, though not with such speed. The female carries its young in a pouch (facing back) like most other quadrapeds of the country.

Western Tasmania   Thomas Scott[102]

21

Captive thylacines   **1863**

The circumstances of a fine pair, male and female . . . now living in Regent's Park Zoo [London], enables me to give the best figure of the animal that has yet appeared . . . Tasmania, better known as Van Diemen's Land, is the country it inhabits, and so strictly is it confined to that island, that I believe no instance is on record of its having been found on the neighbouring continent of Australia. It must be regarded as the most formidable, both of the Marsupialia and of the indigenous mammals of Australia; for although too feeble to make a successful attack on man, it commits sad havoc among the smaller quadrupeds of the country, and among the . . . domestic animals of the settler . . . The destruction it deals around has, as a matter of course, called forth the enmity of the settler, and hence in all cultivated districts the animal is nearly extirpated; on the other hand, so much of Tasmania still remains in a state of nature, and so much of its forest land yet uncleared, that an abundance of cover still remains in which the animal is secure from the attacks of man; many years must therefore elapse before it can become entirely extinct; in these remote districts it preys upon [wallabies], *Bandicoots*, *Echidnae*, and all the smaller animals.

London   **John Gould**[44]

The thylacine (or Tasmanian tiger or wolf) is almost certainly extinct now. No living specimen has been seen since 1933.

When Europeans first arrived, thylacines were abundant in Tasmania; but in the 1840s up to £1 was paid for a scalp because of the threat to the farmers' poultry and sheep. It is thought that after large numbers were killed by hunters, a sudden biological catastrophe, such as disease, lowered numbers so drastically that the population was unable to support itself. Feral dog packs also helped its extinction.

Aboriginal paintings and engravings in mainland Australia depict the thylacine, but when Europeans arrived it was extinct – due, it is thought, to competition with the dingo which the Aboriginals had brought with them centuries before.

A Tasmanian tiger about to enter a 'Tyger Trap' with seeming unconcern. The watercolour is dated 1823 and was painted in a small sketchbook by Thomas Scott.[96]

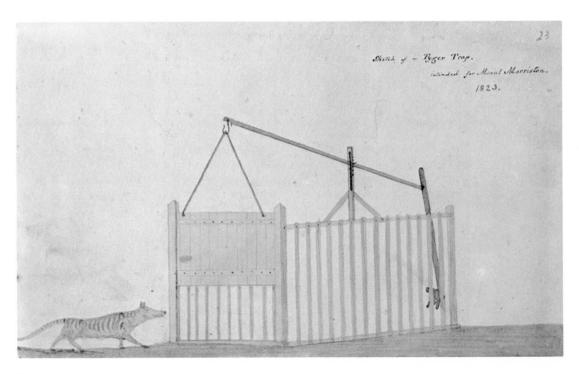

Above: A pair of thylacines by H. C. Richter, who drew many of the pictures in Gould's *Mammals of Australia*[44].

Below: A mother and her three cubs in the Beaumaris Zoo, Hobart – one of the few photographs of the thylacine.[9] The scientific name, *Thylacinus cynocephalus*, means 'pouched dog with the head like a wolf', and certainly the jaws could be opened extraordinarily wide. But it was an attractive animal, and it is our loss that we can no longer see it alive.

Wombat

Vombatus ursinus

The wombat named   **26 January 1798**

We saw several sorts of dung of different animals, one of which Wilson called a Whom-batt, which is an animal about 20 inches high, with short legs and a thick body forwards with a large head, round ears, and very small eyes, is very fat, and has much the appearance of a badger.

Bargo, N.S.W.   **John Price[92]**

First detailed description   **25 August 1798**

It is about the size of a Badger, a species of which we supposed it to be, from its dexterity of burrowing in the earth, by means of its fore paws; but on watching its general motions, it appeared to have much of the habits and manners of the Bear.

Its head is large; the forehead, above the eyes, is particularly broad, from which it tapers to the nose, which is a hard gristly substance, and seems well adapted for removing the earth where it burrows: it has two cutting teeth in each jaw, long and sharp like those of a Kanguroo, with a space of about an inch between them and the grinders, which are strong and well set: from the structure of its teeth, it does not appear to be a carnivorous animal: its eyes are small and black; its ears short and pointed; its paws are somewhat like those of a bear: its weight appeared to be about forty pounds. It runs awkwardly, in the manner of a Bear, so that a man could easily overtake it. There is something uncommon in the form of its hinder parts; its posteriors do not round off like those of most other animals, but fall suddenly down in a sloping direction, commencing at the hip joint, and descending to the knee joint of the hind legs; from this joint to the toe it appears to tread flat upon the ground; its tail is so short, as hardly to be discovered: Its colour is that of a cream-coloured brown, intermixed with black hairs. This animal has lately been discovered to be an inhabitant of the interior of this country also. Its flesh is delicate meat. This one is a female, and has the false belly for the security of its young. The mountain natives call it *Wombach*.

Preservation Island

     Letter from Hunter to Banks[60]

The earliest European illustration of a wombat was made by John Hunter in 1798 and reproduced by Thomas Bewick in 1800.[13] It is curiously elongated.

A friendly wombat 23 June 1808

A male wombat was brought from the islands in Basse's Straits, by Mr. BROWN, the naturalist attached to Captain FLINDERS'S voyage of discovery. It was entrusted to my care, and lived in a domesticated state for two years, which gave me opportunities of attending to its habits.

It burrowed in the ground whenever it had an opportunity, and covered itself in the earth with surprising quickness. It was quiet during the day, but constantly in motion in the night: was very sensible to cold; ate all kinds of vegetables; but was particularly fond of new hay, which it ate stalk by stalk, taking it into its mouth like a beaver, by small bits at a time. It was not wanting in intelligence, and appeared attached to those to whom it was accustomed, and who were kind to it. When it saw them, it would put its fore paws on the knee, and when taken up would sleep in the lap. It allowed children to pull and carry it about, and when it bit them did not appear to do it in anger or with violence.

Bass Strait **Everard Home**[58]

In 1827, Bewick's son engraved a wombat to illustrate a catalogue of the Newcastle Museum, England[38], but he made it sit up like a kangaroo – something it would never do naturally.

The first report of wombats by Europeans was surprisingly late: in February 1797, by sailors stranded by the beaching of the *Sydney Cove* on Preservation Island, off Tasmania. On the mainland, convicts who lived 'up country' showed them to early explorers.

The three wombat species have sometimes been grouped with the possums, but it is probable that they are more closely related to the koala. The common wombat *Vombatus ursinus* is able to live in many types of country, in a burrow protecting it from heat or cold, rain or drought. Wombats remain in the humid air of their burrows most of the day, emerging at night to eat grass, leaves, roots, bark and fungi. The pouch of the female opens backwards and does not fill with soil during digging operations. An adult can weigh up to 35 kilograms; they are Australia's largest burrowing animals.

Koala

Phascolarctos cinereus

The first illustration of a koala, published in Perry's *Arcana* (1811)[89], appears to have been engraved from a stiff, preserved specimen.

First sighting   **26 January 1798**
There is another animal which the natives call a cullawine, which much resembles the sloths in America.
Bargo, N.S.W.   **John Price**[92]

First specimen (part only)   **1802**
[The Aboriginals] brought portions of a monkey (in the native language Colo), but they had cut it in pieces, and the head, which I should have liked to secure, had disappeared. I could only get two feet . . .
I sent these two feet to the Governor preserved in a bottle of brandy.
Locality unknown   **Ensign F. Barrallier**[4]

First published account   **21 August 1803**
An Animal whose species was never before found in the Colony, is in His excellency's possession. When taken it had two Pups, one of which died a few days since – This creature is somewhat larger than the Waumbut, and although it might at first appearance be thought much to resemble it, nevertheless differs from than animal. The fore and hind legs are about of an equal length, having sharp talons at each of the extremities, with which it must have climbed the highest trees with much facility. The fur that covers it is soft and fine, and of a mixed grey colour; the ears are short and open; the graveness of the visage, which differs little in colour from the back, would seem to indicate a more than ordinary portion of animal sagacity; and the teeth resemble those of a rabbit. The surviving Pup generally clings to the back of the mother, or is caressed with a serenity that appears peculiarly characteristic; it has a false belly like the apposim, and its food consists of gum leaves, in the choice of which it is excessively nice.
Port Jackson   **Sydney Gazette**[105]

A soldier's pet   **9 October 1803**
Serjeant Packer of Pitt's Row, has in his possession a native animal some time since described in our Paper, and called by the natives, a Koolah: It has two young, has been caught more than a month, and feeds chiefly on gum leaves, but also eats bread soaked in milk or water.
Port Jackson   **Sydney Gazette**[105]

First scientific report   **23 June 1808**
It is commonly about two feet long and one high, in the girth about one foot and a half; it is covered with fine soft fur, lead coloured on the back, and white on the belly. The ears are short, erect, and pointed; the eyes generally ruminating, sometimes fiery and menacing. It bears no small resemblance to the bear in the fore part of its body; it has no tail; its posture for the most part is sitting.
The New Hollanders eat the flesh of this animal, and therefore readily join in the pursuit of it; they examine with wonderful rapidity and minuteness the branches of the loftiest gum trees; upon discovering the koala, they climb the tree in which it is seen with as much ease and expedition, as an European would mount a tolerably high ladder. Having reached the branches, which are sometimes forty or fifty feet from the ground, they follow the animal to the extremity of a bough, and either kill it with the tomahawk, or take it alive. The koala feeds upon the tender shoots of the blue gum tree, being more particularly fond of this than of any other food; it rests during the day on the

tops of these trees, feeding at its ease, or sleeping. In the night it descends and prowls about, scratching up the ground in search of some particular roots; it seems to creep rather than walk: when incensed or hungry, it utters a long shrill yell and assumes a fierce and menancing look. They are found in pairs, and the young is carried by the mother on its shoulders.

S.W. of Port Jackson   **Everard Home**[58]

The koala feet were pickled at considerable cost to Barrallier, for brandy was then scarce in Sydney Town.

The koala inhabits only the eastern Australian mainland. Although it is usually found in trees, it can run surprisingly fast and, when forced, can swim. Each adult eats more than 1 kilogram of gum leaves a day, but only from a few selected species of eucalypts. The pouch of the female opens backwards, enabling the young koala to reach its mother's anus. At weaning, the mother produces, and the young eats, a special kind of faeces which contain micro-organisms necessary for gum leaf digestion. Like possums, the koala can be quite vocal – whining and growling at its friends and enemies.

It used to be trapped for its fur – 600 000 skins were marketed in 1927 – but it is now totally protected, and numbers are slowly increasing. Koalas carry a fungal disease, which may affect people fatally.

The koala is not a bear, although the biologist who named it was less accurate than he could have been. Its scientific name means 'ashy-coloured pouched bear', from *phascolos* a bag or wallet, *arctos* a bear, and *cinereus* ashy-coloured.

Gould's female koala and baby, drawn from life, sit at ease in their gum trees. Note the separation of the fingers and toes – a scientifically accurate detail which earlier observers missed.[44]

Wallabies (small macropods)

First sighting of the Tammar wallaby

15 November 1629

Besides we found in these islands large numbers of a species of cats, which are very strange creatures; they are about the size of a hare, their head resembling the head of a civet-cat; the forepaws are very short, about the length of a finger, on which the animal has five small nails or fingers, resembling those of a monkey's forepaw. Its two hind legs, on the contrary, are upwards of half an ell in length [about 500 mm] and it walks on these only, on the flat of the heavy part of the leg, so that it does not run fast. Its tail is very long, like that of a long-tailed monkey; when it eats, it sits on its hind legs, and clutches its food with its forepaws, just like a squirrel or monkey. Their manner of generation or procreation is exceedingly strange and highly worth observing. Below the belly the female carries a pouch, into which you may put your hand; inside this pouch are her nipples, and we have found that the young ones grow up in this pouch with the nipples in their mouths. We have seen some young ones lying there, which were only the size of a bean, though at the same time perfectly proportioned, so that it seems certain that they grow there out of the nipples of the mammae, from which they draw their food, until they are grown up and able to walk. Still, they keep creeping into the pouch, even when they have become very large, and the dam runs off with them, when they are hunted.

Abrolhos Islands, W.A.   **Francis Pelsaert**[112]

First sighting of the quokka

1656

In slightly under 32 – S. latitude there is a large island, at about 3 miles' distance from the mainland of the South-land . . . here certain animals are found, since we saw many excrements, and besides two seals and a wild cat, resembling a civet-cat, but with browner hair.

Rottnest Island, W.A.   **Samuel Volkertszoon**[56]

First sighting of the banded hare-wallaby

6 August 1699

The Land-Animals that we saw here were only a Sort of Raccoons, different from those of the West-Indies, chiefly as to their Legs; for these have very short Forelegs; but go jumping upon them as the others do (and like them are very good Meat).

Dirk Hartog Island, W.A.   **William Dampier**[30]

Wallabies were the first kangaroo-type animals to be seen by Europeans. Pelsaert's description is amazingly accurate for a first sighting, but his conjecture that the young grow out of the nipples was a misconception that persisted for two hundred years. We now know that marsupials are born like other mammals (excepting monotremes) but in a very immature state, which requires that they be kept in the pouch in a kind of natural humidicrib for some time.

The wallaby's tail is used for balance, both while standing and while bounding through the bush.

This may be the first European picture of a member of the macropod group. It is from the title page of Cornelis de Jode's *Speculum Orbis Terrae*, published in Antwerp in 1593.[33]

The first illustration recognisable as a wallaby was published in 1714 in a book of travels (de Bruin[32]). It was drawn in Java from a captive wallaby, the Filander *Dorcopsis brunii* from New Guinea.

Kangaroos (large macropods)

This quaint illustration appeared in 1802 in George Barrington's *History of New South Wales*.[5] It incorrectly portrays a kangaroo picking oak-like leaves from a tree and offering them to her young. Twins are rare, although as soon as one joey vacates the pouch a second is often born, and the mother may therefore suckle two offspring at one time.

First sighting of a kangaroo   22 June 1770

The People who were sent to the other side of the water in order to shoot Pigeons saw an animal as large as a grey hound, of a mouse coulour and very swift, they also saw many Indian houses and a brook of fresh water.

Endeavour River, Qld   **Joseph Banks**[8]

Cook sees a kangaroo   24 June 1770

I saw my self this morning a little way from the ship one of the Animals before spoke off, it was of a light Mouse colour and the full size of a grey hound and shaped in every respect like one, with a longtail which it carried like a grey hound, in short I should have taken it for a wild dog but for its walking or runing in which it jumped like a Hare or a dear; Another of them was seen to day by some of our people who saw the first, they described them as having very small legs and the print of the foot like that of a goat, but this I could not see my self because the ground the one I saw was upon was too hard and the length of the grass hinderd my seeing its legs.

Endeavour River, Qld   **James Cook**[7]

. . . and so does Banks   25 June 1770

In gathering plants today I myself had the good fortune to see the beast so much talkd of, tho but imperfectly; he was not only like a grey hound in size and running but had a long tail, as long as any grey hounds; what to liken him to I could not tell, nothing certainly that I have seen at all resembles him.

Endeavour River, Qld   **Joseph Banks**[8]

First kangaroo shot   14 July 1770

Mr. Gore being out in the Country shott one of the Animals before spoke of, it was a small one of the sort weighing only 28 pounds clear of the entrails. [This was probably an eastern wallaroo.] The head neck and shoulders of this Animal was very small in proportion to the other part; the tail was nearly as long as the body, thick next the rump and tapering towards the end; the fore legs were 8 Inch long and the hind 22, its progression is by hoping or jumping 7 or 8 feet at each hop upon its hind legs only, for in this it makes no use of the fore, which seem to be only design'd for scratching in the ground and etc. The Skin is cover'd with a short hairy fur of a dark Mouse or Grey Colour. Excepting the head and ears which I thought was something like a Hare's, it bears no sort of resemblance to any European Animal I ever saw; it is said to bear much resemblance to the Gerbua excepting in size, the Gerbua being no larger than a common rat.

Endeavour River, Qld   **James Cook**[7]

. . . and eaten   15 July 1770

The Beast which was killd yesterday was today Dressd for our dinners and provd excellent meat.

Endeavour River, Qld   **Joseph Banks**[8]

First grey kangaroo shot   27 July 1770

This day was dedicated to hunting wild animal. We saw several and had the good fortune to kill a very large one which weighd 84 lb.

Endeavour River, Qld   **Joseph Banks**[8]

The animal named   4 August 1770

. . . the Animal which I have before mentioned is called by the natives *Kangooroo* or *Kanguru*.

Endeavour River, Qld   **James Cook**[7]

James Cook's *Endeavour* was holed by a piece of coral in June 1770. While beached for repairs (near what was named the Endeavour River, in Queensland), Cook saw his first kangaroo; he was just as excited as the many other Europeans who had seen a kangaroo before him – de Bruin, Pelsaert, Volkertszoon, Vlamingh, Dampier and others, including several of his own party.

Kangaroos have several adaptations to live in dry conditions: they do not sweat; if too hot they lick their chest and arms; they have nostrils that can be closed; the upper lip is cleft from the opening of the nostrils to the mouth, so that any moisture escaping from the nose runs to the mouth; the entrances to the ears and eyes are protected by stiff hairs; and the long roof of the mouth can act as an efficient evaporative cooler.

The second and third toes of kangaroos are joined together part-way along their length and act as a grooming comb; this adaptation is also found in possums, koalas, bandicoots and wombats.

Kangaroos are vegetarians, eating grass or leaves. In general the larger kangaroos have benefited from the European invasion – having more grass and water than previously; but much of the undergrowth in which the smaller species live has been destroyed, and so they have become scarce.

THE WONDERFUL
KANGUROO,

FROM
BOTANY BAY,

(The only One ever brought alive to Europe)

Removed from the HAY-MARKET, and now exhibited at the LYCEUM,
in the STRAND, from 8 o'Clock in the Morning, till 8 in the Evening.

THIS amazing, beautiful, and tame Animal, is about five Feet in Height, of a Fawn
Colour, and diftinguifhes itfelf in Shape, Make, and true Symmetry of Parts,
different from all other QUADRUPEDS. Its Swiftnefs, when purfued, is fuperior to
the Greyhound: to enumerate its extraordinary Qualities would far exceed the common
Limits of a Public Notice. Let it fuffice to obferve, that the Public in general are
pleafed, and beftow their Plaudits; the Ingenious are delighted; the Virtuofo, and
Connoiffeur, are taught to admire! impreffing the Beholder with Wonder and Afto-
nifhment, at the Sight of this unparalleled Animal from the Southern Hemifphere, that
almoft furpaffes Belief; therefore Ocular Demonftration will exceed all that Words
can defcribe, or Pencil delineate........Admittance, ONE SHILLING each.

This early handbill, now in the Mitchell Library, Sydney, captures some of the astonishment of Europeans when first they saw a kangaroo.

Some biological notes   **1788**

Of the natural history of the kangaroo we are still very ignorant. We may, however, venture to pronounce this animal, a new species of opossum, the female being furnished with a bag, in which the young is contained; and in which the teats are found. These last are only two in number, a strong presumptive proof, had we no other evidence, that the kangaroo brings forth rarely more than one at birth. But this is settled beyond a doubt, from more than a dozen females have been killed, which had invariably but one formed in the pouch. Notwithstanding this, the animal may be looked on as prolific, from the early age it begins to breed at, kangaroos with young having been taken of not more than thirty pounds weight; and there is room to believe that when at their utmost growth,

they weigh not less than one hundred and fifth pounds. A male of one hundred and thirty pounds weight has been killed, whose dimensions were as follows:

	Ft.	Inch.
Extreme length	7	3
Do. of the tail	3	4½
Do. of the hinder legs	3	2
Do. of the fore paws	1	7½
Circumference of the tail at the root	1	5

After this perhaps I shall hardly be credited, when I affirm that the kangaroo on being brought forth is not larger than an English mouse. It is, however, in my power to speak positively on this head, as I have seen more than one instance of it.

In running, this animal confines himself entirely to his hinder legs, which are possessed with an extraordinary muscular power. Their speed is very great, though not in general quite equal to that of a greyhound; but when the greyhounds are so fortunate

32

as to seize them, they are incapable of retaining their hold, from the amazing struggles of the animal. The bound of the kangaroo, when not hard pressed, has been measured, and found to exceed twenty feet.

At what time of the year they copulate, and in what manner, we know not: the testicles of the male are placed contrary to the usual order of nature.

When young the kangaroo eats tender and well flavoured, tasting like veal, but the old ones are more tough and stringy than bull-beef. They are not carnivorous, and subsist altogether on particular flowers and grass. Their bleat is mournful, and very different from that of any other animal: it is, however, seldom heard but in the young-ones.

New South Wales   **Watkin Tench**[110]

Naturalised in England   **1805**

The Kanguroo may now be considered as in a great degree naturalized in England; several having been kept for some years in the royal domains at Richmond, which, during their residence there, have produced young, and apparently promise to render this most elegant animal a permanent acquisition to our country; though it must, no doubt, lose, by confinement and alteration of food, several of its natural habits, and exhibit somewhat less of that bounding vivacity which so much distinguishes it in its native wilds of New Holland.

Surrey, England   **Rev. W. Bingley**[14]

'Escaped Kangaroo at Regent's Park', London, by R. Cruikshanks, *c*. 1840.[18]

'Kangaroo Hunt, Mount Zero, The Grampians, Victoria', Edward Roper, 1880 (oil).[93]

Short-nosed bandicoot

Isoodon obesulus

The early sightings of Australian animals were often imprecise, and Dr Solander can hardly be blamed for describing the bandicoot as something like a rabbit. (Rabbits were landed with the First Fleet in 1788, although most in Australia today are descended from twenty-four wild rabbits released in Victoria in 1859.)

The first European settlers must soon have noticed the short-nosed bandicoot *Isoodon obesula* and the long-nosed bandicoot *Perameles nasuta*; like the brush-tailed possum, they adapted well to towns and cities.

Bandicoots are fossickers; they scrabble and dig for insects, worms and snails in the soil and leaf litter, digging small conical holes in suburban gardens. Their name is Indian, meaning 'pig-rat' — they look superficially similar to some Indian rats. Like the koala, bandicoots rarely drink, obtaining the water they need from dew and their food.

First sighting?   **1 May 1770**

Dr Solander had a bad sight of a small Animal something like a rabbit . . .

Botany Bay   **James Cook**[7]

The species described   **1790**

This species hitherto undescribed, is a native of New Holland and is remarkable for a thicker or more corpulent habit than others of the genus. The hind legs are considerably longer than the fore legs and have in miniature the form of those of the Kangaroo and some other Australasian quadrupeds, though the middle claws are far less in proportion. The interior [claws] are double or both covered by a common skin. The size of this species is nearly that of a small or half grown domestic rat, its colour is pale yellow-brown and its hair is of a coarser or more hard appearance than in the rest of the smaller opposums [*sic*]; the ears are rounded; the tail rather long. When viewed in a

cursory manner, the animal bears a distant resemblance of a pig in miniature.
Botany Bay   **George Shaw**[97]

Observations on nests   1863
While engaged in my observations on the 'Birds of Australia', I have very frequently trodden upon the almost invisible nest of this species and aroused the sleeping pair within, which would then dart away with the utmost rapidity, and seek safety in the dense scrub, beneath a stone, or in the hollow bole of a tree . . . The following note is from the pen of the late Mr. Gilbert . . . 'This little animal is abundant in every part of the colony, and is found in every variety of situation; in thick scrubby places, among the high grass growing along the banks of rivers and swamps, and also among the dense underwood both on dry elevated land and in moist situations. It makes a nest of short pieces of dried sticks, coarse grasses, leaves, etc. sometimes mixed with earth, and so artfully contrived to resemble the surrounding ground, that only an experienced eye can detect it. When built in dry places, the top is flat, and on a level with the ground, but in moist situations the nest is often raised in the form of a heap, to the height of about twelve inches; the means of access and exit being most adroitly closed by the animal both on entering and emerging . . .'
Australia   **John Gould**[44]

Published in 1846 in G. R. Waterhouse's *Natural History of the Mammalia*[114], this illustration is more accurate than most earlier ones, which tended to make the bandicoots too rat-like. This specimen, though, is rather too fat and still gives a poor idea of these nimble animals.

These short-nosed bandicoots are from Waterhouse's *Natural History of the Marsupialia or Pouched Animals*.[113] Even a short-nosed bandicoot has a nose of respectable length. Including the tail, an adult measures about 350 mm.

Possums

Possums exist in considerable variety in Australia and exact identification cannot be made from the early reports. Today the word 'opossum' is used to refer to the American marsupials, and 'possum' to the Australian, of which there are about 24 species. Sizes vary, including the tail, from 150 to 900 mm.

Some possums build nests or 'dreys' in which they sleep at night. Most eat their own faeces, apparently ensuring that undigested and unabsorbed vitamins and minerals are not wasted. The common brushtail *Trichosurus vulpecula* has adapted well to European invasion of Australia and is common – even a pest – in cities; it is easily tamed.

The early explorers were quick to note the prehensile tail of some species and the specially adapted climbing hands of possums.

First sighting   **26 August 1770**

In botanizing today I had the good fortune to take an animal of the Opossum (*Didelphis*) tribe: it was a female and with it I took two young ones. It was not unlike that remarkable one which De Bufon has described by the name of Phalanger as an American animal; it was however not the same for De Buffon is certainly wrong in asserting that this tribe is peculiar to America; and in all probability, as Pallas has said in his *Zoologia*, the Phalanger itself is a native of the East Indies, as my animal and that agree in the extraordinary conformation of their feet in which particular they differ from all the others.

Endeavour River, Qld   **Joseph Banks[8]**

Further observations   **30 January 1777**

The only animal of the Quadruped kind we got was a sort of Opossum about twice the size of a large cat and liker that than any other creature. It is of a dusky colour above tingd with a brown or rusty cast and below it is whitish: about a third of the tail towards its tip is white and bare underneath, by which it probably hangs on the branches of trees as it climbs these and lives on berrys. For this purpose the feet are made much like a little hand and the hind ones has a thumb distinct from the others but without a claw.

Adventure Bay, Tas.   **William Anderson[7]**

A plate from George Shaw's *General Zoology*, published in 1800.[100] The animal at the top is, in fact, a marsupial mouse, not a possum.

A gliding possum   **5 February 1809**

A curious little native Animal, partaking in appearance of the squirrel & native cat is at this time in possession of Dan. M'Coy. The tail is of a remarkable length, the eyes rather prominent, the breast of a deep tan-colour, and the back a dark mixture. On either shoulder, a thin skin expands itself, which probably assisted in its flight or spring from tree to tree. From its hoarse droning noise when handled or molested, it at present passes under the appellation of the Buzzing Squirrel.

Sydney   **Sydney Gazette[107]**

This gliding possum with outspread limbs is one of many hand-coloured engravings of animals in John White's *Journal of a Voyage to New South Wales* (1790).[115]

Feathertailed glider

Acrobates pygmaeus

Gliding possums part jump, part parachute and part glide from a higher branch to a lower. They have a fold of skin between the arm and leg on each side of the body, which they outstretch when they launch themselves. The tail is long and flattened; it helps to increase air resistance and maintain balance. The largest gliders can jump more than 80 metres, although the tiny feathertailed glider cannot go as far.

Gliding possums are found over much of coastal eastern Australia. They eat blossoms, fruit and insects, and like other possums are quite vocal. The feathertail is probably commoner than it seems to be because of its secretive habits.

The feathertailed glider is only about 150 mm long, including the tail. This specimen in George Waterhouse's *Marsupialia* (1841)[113] appears to be of immense size because of the disproportionately small plants around it.

First description of the 'pygmy opossum' 1793
Amongst the most curious quadrupeds yet discovered in the Antarctic regions, may be numbered the animal represented on the present plate; which (exclusive of its diminutive size, not exceeding that of a common domestic mouse) forms as it were a kind of connecting link between the genera of . . . Opossum and Squirrel . . . In the present instance, however, I have not disassociated this species from the other Didelphides; and as it is probably by far the most diminutive of the tribe, have distinguished it by a title expressive of its smallness.

New Holland   **George Shaw**[98]

Early description   **1841**
Ears rather small . . . fur soft, and of moderate length; general colour grey, more or less washed with yellow . . . a dark patch surrounds the eye, and is extended forward on to the muzzle; the under parts of the head and body are white; the tail is brown, nearly equal in length to the head and body, flat, of even width throughout, or very nearly so . . . Habitat, New South Wales. It is called by the Colonists *the flying mouse*, and, according to the Appendix to Captain King's *Narrative of a Survey of The Intertropical and Western coasts of Australia*, it is "exceedingly numerous in the vicinity of Port Jackson", where it is called the Opossum Mouse.

New South Wales   **G. R. Waterhouse**[113]

The 'opossum mouse'   **1863**
This pretty little animal, the "Opossum Mouse" of the colonists, is very common in every part of New South Wales; but from its nocturnal habits, its small size, and from the circumstance of its exclusively inhabiting the hollow limbs of the larger gum-trees, it rarely comes under the observation of ordinary travellers; it is in fact seen in considerable numbers only by those who really live in the bush, and to their notice it is seldom presented except under extraordinary circumstances, the most frequent of which are the blowing off of a large limb in which it is concealed; but if, as occurred several times during my explorations, the limb be thrown upon the traveller's fire, the little inhabitant is soon driven forth by the heat: occasionally as many as four or five are discovered by this means . . . a more charming little pet cannot be imagined, an ordinary-sized pill-box forming a convenient domicile for the tiny creature, in which it

lies coiled up during the day, becoming more and more active as night approaches. Its food consists of the saccharine matter which is so abundant in the flower-cups of the ever-blossoming *Eucalypti*, for which well-sweetened bread and milk forms an excellent substitute.

New South Wales   **John Gould**[44]

J. Sowerby's engraving accompanied the first description. It appeared in Shaw's *Zoology and Botany of New Holland*[98], published in 1793; and shows one pygmy possum licking sugary sap, the other gliding with outstretched membranes.

Flying foxes

Pteropus spp.

Bats are the only mammals that can fly. Their bodies are furred, the wings bare skin. There are two main kinds of bats: Microchiroptera ('little hand-winged' animals), which are usually small, insectivorous and cave- or hole-dwelling; and Megachiroptera ('large hand-winged' animals), which are usually larger and fruit- or blossom-eating and tree-hanging.

The species of the Megachiroptera in Australia are called flying foxes (*Pteropus* spp.). They gather in large flocks, following seasonal ripening of fruit. On becoming airborne they seem to increase suddenly in size, for the wing span is four times the length of their bodies (260 mm); they look black against the sky, although in some species the fur is quite orange-coloured.

The baby bat clings to its mother's fur during the first month of life, before learning to fly.

A masterpiece of sophisticated observation, drawn by H. C. Richter[44] from skins supplied from the H.M.S. *Rattlesnake* expedition. All bats have wings formed by a membrane being stretched over the four very elongated fingers; the short thumb has a well-developed claw to aid in climbing.

First sighting   24 June 1770

A seaman who had been out in the woods brought home the description of an animal he had seen composd in so Seamanlike a stile that I cannot help mentioning it: it was (says he) about as large and much like a one gallon keg, as black as the Devil and had 2 horns on its head, it went but Slowly but I dard not touch it.

Endeavour River, Qld   **Joseph Banks**[8]

Food for explorers   November 1844

Myriads of flying-foxes were here suspended in thick clusters on the highest trees in the most shady and rather moist parts of the valley. They started as we passed, and the flapping of their large membranous wings produced a sound like that of a hail-storm . . . I went with Charley and Brown to the spot where we had seen the greatest number of flying-foxes, and, whilst I was examining the neighbouring trees, my companions shot sixty-seven, of which fifty-five were brought to our camp; which served for dinner, breadfast, and luncheon, each individual receiving eight.

Queensland   **Ludwig Leichhardt**[71]

A new species   1846

A new species of large fruit-eating bat, or "flying-fox", (*Pteropus conspicillatus*), making the third Australian member of the genus, was discovered here. On the wooded slope of a hill I one day fell in with this bat in prodigious numbers, presenting the appearance, while flying along in the bright sunshine, so unusual in a nocturnal animal, of a large flock of rooks. On close approach a strong musky odour became apparent, and a loud incessant chattering was heard. Many of the branches were bending under their loads of bats, some in a state of inactivity, suspended by their hind claws, others scrambling along among the boughs, and taking to wing when disturbed. In a very short time I procured as many specimens as I wished, three or four at a shot, for they hung in clusters, – but, unless killed outright, they remained suspended for some time, – when wounded they are to be handled with difficulty, as they bite severely, and on such occasions their cry reminds one of the squalling of a child. The flesh of these large bats is reported excellent; it is a favourite food with the natives, and more than once furnished a welcome meal to Leichhardt and his little party, during their adventurous journey to Port Essington.

Queensland   **John MacGillivray**[77]

A macabre engraving from William Dampier's journals.[30] This particular specimen was seen in New Guinea, rather than in Australia.

Dingo
Canis familiaris

First sighting   **8 May 1623**
I went ashore myself with 10 musketeers; we saw numerous footprints of men and dogs (running from south to north); we accordingly spent some time there, following the footprints to a river . . . we also saw great numbers of dogs, herons and curlews, and other wild fowl, together with plenty of excellent fish.
Near Cape Keer Weer, Qld   **Jan Carstenzoon**[27]

An energetic dingo   **January 1697**
Some others said . . . they saw a yellow dog leaping from the wild herbage, and throwing itself into the sea, as if to amuse himself with swimming.
Inland from Jurien Bay, W.A.
                      William de Vlamingh[56]

Another report   **October 1698**
On this Voyage nothing hath been discovered which can be any way serviceable to the Company. The Soil of this Country hath been found very barren, and as a Desart; no Fresh-water Rivers have been found, but some Salt-water Rivers, as also no Four-footed Beasts, except one as great as a Dog, with long Ears, living in the Water as well as on the Land.
Jurien Bay, W.A.   **Nicolaus Witsen**[74]

The dingo was frequently illustrated in early colonial accounts; this picture is from Barrington's *History of New South Wales* (1802).[5]

An early settler's account   **1802**
The native dog of New South Wales resembles very much the foundation of the species, which is the Wolf, though it is considerably smaller, and stands lower; but from its ill-nature and viciousness, which indeed nothing overcomes, it may with great propriety be esteemed the wolf of the country.

    The Dog or Dingo barks in a way peculiar to itself, but moans, snarls, and howls like other dogs.

    Its general colour is a reddish dun, covered with long thick stait hair, and has short erect ears and a bushy tail; the nose, belly and feet are of a blue grey colour.
New South Wales   **George Barrington**[5]

Ability to endure pain   **1834**

The cunning displayed by these animals, and the agony they can endure without evincing the usual effects of pain, would seem almost incredible, had it not been related by those on whose testimony every dependence can be placed. The following are a few among a number of extraordinary instances. One had been beaten so severely, that it was supposed all the bones were broken, and it was left for dead. Upon the person accidentally looking back, after having walked some distance, his surprise was much excited by seeing 'master dingo rise, shake himself, and march into the bush, evading all pursuit'. One supposed to be dead was brought into a hut, for the purpose of undergoing 'decortication', – at the commencement of the skinning process upon the face, the only perceptible movement was a slight quivering of the lips, which was regarded at the time as merely muscular irritability: the man, after skinning a very small portion, left the hut to sharpen his knife, and returning, found the animal sitting up, with the flayed integument hanging over one side of the face.

New South Wales   **George Bennett**[10]

The dingo was brought to Australia by the Aboriginals, and is descended from the plains wolf of Asia. It can mate with the domestic dog and cross-breeds are common.

Europeans saw the dingo early in their explorations and noted the yellow colour of its coat (Witsen was with de Vlamingh in 1697-98 but was mistaken about its living in water); later reports describe its strength. Since the introduction of sheep into Australia, the dingo has varied its previous diet of small kangaroos and rats.

Gould's painting shows four dingoes in various states of activity[44]; note the muscular body. When full grown, the dingo is 1400 mm long.

Dugong

Dugong dugon

The dugong is a mammal which lives in the sea, like seals, whales and dolphins; but whereas they are carnivorous, the dugong eats marine plants. It is a sea-cow, related to the manatees of Africa and America.

A baby is born underwater and immediately taken to the surface, where it rides on its mother's back for several hours, until it learns to swim. It is suckled under water, though the mother ensures that it breathes at the surface periodically.

It inhabits tropical waters, with others in a small group, and adults can weigh up to 250 kilograms. It has few natural enemies, except sharks and man; Aboriginals regarded the dugong as a useful source of food, though few are killed by them today.

First sighting   **5 January 1688**
Neither is the Sea very plentifully stored with Fish, unless you reckon the Manatee and Turtle as such. Of these Creatures there is plenty; but they are extraordinary shy; though the Inhabitants cannot trouble them much, having neither Boats nor Iron.
Near King Sound, W.A.   William Dampier[30]

A good food   12 January 1688
Most of our Men lay ashore in a Tent, where our Sails were mending; and our Strikers brought home Turtle and Manatee every day, which was our constant Food.
Near King Sound, W.A.   William Dampier[30]

A curious capture   6 August 1699
Of the Sharks we caught a great many, which our Men eat very savourily. Among them we caught one which was 11 Foot long. The space between its two Eyes was 20 Inches, and 18 Inches from one Corner of his Mouth to the other. Its Maw was like a Leather Sack, very thick, and so tough that a sharp Knife could scarce cut it: In which we found the Head and Boans of a Hippopotomus [a partly digested dugong]; the hairy Lips of which were still sound and not putrified, and the Jaw was also firm, out of which we pluckt a great many Teeth, 2 of them 8 Inches long, and as big as a Man's Thumb, small at one end, and a little crooked; the rest not above half so long. The Maw was full of Jelly which stank extreamly: However I saved for a while the Teeth and the Sharks Jaw: The Flesh of it was divided among my Men; and they took care that no waste should be made of it.
Shark Bay, W.A.   William Dampier[30]

A sea cow   12 September 1792
Since we have been among these islands and reefs we have not seen any fish common to the sea, but we have seen a number of large porpoises, many of them white and others piebald, and a large brownish kind of animal who make and colour lead me to believe they are the sea cow.
Torres Strait   Nathaniel Portlock[91]

This nineteenth-century engraving shows the horizontally flattened tail of the dugong and exaggerated nose; in the background is a detail meant to be a dugong expelling air, incorrectly after the fashion of a whale.[64]

Birds

There are now almost 800 bird species described from Australia; one of the last ones found was the Eungella honeyeater, described in 1979. From this array, we have chosen twenty-five, which although a large proportion of the animals in the book is but a small selection from the total number possible. Before the First Fleet arrived in 1788, there were already ninety-three species of bird relating to Australia described or at least referred to in journals.

One of the most notable birds is the emu, second largest of the world's flightless birds, growing to 2 metres in height. It is related to other primitive flightless birds, such as ostriches, rheas and kiwis, all of which are found in the southern hemisphere. At one time these birds flew, but gradually the advantages of flight were 'traded' for the advantages of increased body size.

The only stork found in Australia is the jabiru, its range extending through South-east Asia to India. It was once quite plentiful in New South Wales, but largely because of the drainage of swamps for agricultural purposes it is now uncommon throughout the south of Australia.

Early European observers, familiar with white swans only, found the black swan remarkable; and the Cape Barren goose, which they originally thought of as a juvenile black swan, then named a goose, is in fact not a true goose at all. Ornithologists now believe it to be a primitive link between the ducks and the geese.

The megapodes are an unusual group of fowl-like birds confined to the Australasian region (Australia, Papua New Guinea, New Zealand and associated islands); the three Australian species are the brush turkey, the scrub fowl and the mallee fowl. All incubate their eggs with heat generated either from rotting vegetation (the brush turkey and the scrub fowl) or from sun-heated sand (the desert-dwelling mallee fowl). The incubation mounds may be up to 15 metres across by 4 metres high, when used over many years, and there are reports of even larger mounds. To maintain the right temperature, material is continually added or removed by the birds' large feet. When the chicks hatch, they dig themselves from the mound and are immediately able to fly and look after themselves.

Changes to the natural habitat of many species have threatened several with extinction. The Lord Howe Island woodhen was in 1788 common enough to be collected by hand in large numbers, but now its population is reduced to only about thirty. A similar though less dramatic decline in numbers has occurred with the bustard; this species is now common only in areas well away from human activities. The accounts reproduced here show that it was appreciated mainly as a food, as is still the case today, despite the bird's protected status.

There are some sixty species of Australian parrot, and many are common enough to be a prominent feature in many parts of the country. Early European visitors often commented on the 'beautiful loryquets and parroquets', and one map even had Australia shown as *Terra*

44

Psittacorum — the land of parrots. There are two species of cockatoo included here: the sulphur-crested cockatoo and the red-tailed black cockatoo. The former, known to many people as a caged pet, occurs in large flocks in inland forests. Certain species of cockatoo extend into Indonesia; however, black cockatoos are confined to Australia and New Guinea.

The rosellas, another common group of parrots, are represented here by the crimson rosella, the first of the group to be discovered, and the eastern rosella. Lories and lorikeets are a unique group of nectar, flower and pollen feeding parrots. The bright feathers and shrill calls of the rainbow lorikeet attracted immediate attention on Cook's first voyage in 1770; the one illustrated here is notable because it was the first eastern Australian bird taken alive to Europe.

The budgerigar, the world's most popular cage bird and best-known parrot, is found in many colour mutations. The cage birds all had their origin in the little green and yellow grass parrakeet which is still seen today in immense flocks in the inland. Contrasting with this species, the paradise parrot is so rare that it is believed to be extinct; its disappearance is attributed to the loss of seeding grasses caused by severe droughts and widespread overgrazing by cattle. It was described as 'the most beautiful parrot yet found in Australia'.

Lyrebirds are the world's largest perching birds, the two species occurring only in Australia. As well as their beautiful tail feathers, they are noted for unusual anatomical features, such as the structure of the voicebox and associated musculature.

Honeyeaters are a group that extends all over the Australasian region. In Australia there are some seventy species, which have evolved to take up many different ways of life. Thus there are spinebills, which feed on long tubular flowers with their long curved bills; *Melithreptus* honeyeaters, which eat leaf insects more than nectar; and large wattlebirds and friar birds. The two species of honeyeater in the book are the New Holland honeyeater and the yellow-faced honeyeater.

The group containing bowerbirds and birds of paradise is another Australasian family, although some species extend into Indonesia. The bowerbirds were a source of wonder to early naturalists: John Gould rated his description of bower building in these birds as his greatest ornithological achievement. There are nine species of bowerbird in Australia, but accounts recorded here refer only to the great bowerbird, the fawn-breasted bowerbird and the regent bird.

Emu

Dromaius novaehollandiae

Although common in Australia – featuring with the kangaroo on the Australian coat of arms – this large flightless bird is the sole representative of an ancient family found nowhere else in the world. Its scientific name means swift-footed bird of New Holland.

The emu eats leaves, grasses, fruits, flowers, seeds and insects. During incubation of some six to eight weeks, the male warms the eggs with his breast.

Tracks first seen　　　　**15 January 1697**
We had nearly proceeded a league and a half inland; but we saw no men nor fresh water, but several footsteps of men, and steps like those of a dog and of the cassowary.
Jurien Bay, W.A.　　　**Willem de Vlamingh**[56]

First sighting　　　　　**21 January 1788**
The animals we saw during our stay in New Holland . . . were Kongoroos, about as big as a large sheep, a very large species of lizard, dogs, rats, racoons, flying squirrels – very large snakes – a bird of a new genus, as large and high as an Ostritch.
Port Jackson　　　　　**Arthur Bowes**[19]

First shot and clearly described　February 1788
A New Holland Cassowary was brought into camp. This bird stands seven feet high, measuring from the ground to the upper part of the head, and, in every respect, is much larger than the common Cassowary of all authors, and differs so much therefrom, in its form, as to clearly prove it a new species. The colour of the plumage is greatly similar, consisting of a mixture of dirty brown and grey; on the belly it was somewhat whiter; and the remarkable structure of the feathers, in having two quills with their webs arising out of one shaft, is seen in this as well as the common sort. It differs materially in wanting the horny appendage on the top of the head. The head and beak are much more like those of the ostrich than the common Cassowary, both in shape and size.
Two miles from Sydney Cove　**John White**[115]

The first European depiction of an emu, drawn by surgeon Arthur Bowes in the diary he wrote on the transport ship *Lady Penrhyn* in 1788.[19]

First eggs and young　　　　　　　**1793**
I have nevertheless had the good fortune to see what was never seen but once, in the country I am describing, by Europeans – a hatch, or flock of young cassowaries, with the old bird. I counted ten, but others said there were twelve. We came suddenly upon them, and they ran up a hill, exactly like a flock of turkies, but so fast that we could not get a shot at them. The largest cassowary ever killed in the settlement weighed ninety-four pounds: three young ones, which had been by accident separated from the dam, were once taken, and presented to the governor. They were not larger than so many pullets, although at first sight they appeared to be so, from the length of their necks and legs. They were very beautifully striped, and from their tender state, were judged to be not more than three or four days old. They lived only a few days.

Emu feathers appear to grow in pairs, as shown in this illustration from Captain Arthur Phillip's *Voyage to Botany Bay* (published 1789).[90] The pairing is due to a lengthening of the aftershaft of the feather. The bird was drawn by Lieutenant Watts.

An illustration of the emu which appeared in White's journal.[115] Note the over-prominent wings.

A single egg, the production of a cassowary, was picked up in a desart place, dropped on the sand, without covering or protection of any kind. Its form was nearly a perfect ellipsis; and the colour of the shell a dark green, full of little indents on its surface. It measured eleven inches and a half in circumference, five inches and a quarter in height, and weighed a pound and a quarter. – Afterwards we had the good fortune to take a nest: it was found by a soldier, in a sequestered solitary situation, made in a patch of lofty fern, about three feet in diameter, rather of an oblong shape, and composed of dry leaves and tops of fern stalks, very inartificially put together. The hollow in which lay the eggs, twelve in number, seemed made solely by the pressure of the bird . . . The soldier, instead of greedily plundering his prize, communicated the discovery to an officer who immediately set out for the spot. When they had arrived there, they continued for a long time to search in vain for their object; and the soldier was just about to be stigmatised with ignorance, credulity, or imposture, when suddenly up started the old bird, and the treasure was found at their feet. The food of the cassowary is either grass, or a yellow bell-flower growing in the swamps. – It deserves remark, that the natives deny the cassowary to be a bird, because it does not fly.

Port Jackson   **Watkin Tench**[109]

Black swan

Cygnus atratus

The black swan, symbol of the state of Western Australia and pictured on some of Australia's most valuable stamps, was one of the first Australian birds to be recorded by European visitors. Distributed over the whole of the southern half of the continent wherever there is suitable water, black swans may be seen in flocks, feeding on water plants and grazing on pasture adjacent to waterways.

First sighting   **5 July 1636**
An entry in the ship's log reports the sighting of two large black birds with orange-yellow bills almost half a yard long.
Bernier Island, N.W. Australia   **Antonie Caen**[23]

First capture   **7 January 1697**
We found two young swans on the river which we overtook by rowing fast, caught them with a hook which slightly injured one of them whereas the other was not hurt, and brought them aboard. They are quite black.
Jurien Bay, W.A.   **Willem de Vlamingh**[112]

Willem de Vlamingh's men in chase of black swans: a woodcut of the scene in January 1697 when this species was first caught by Europeans – on the Swan River, W.A.[112]

A miracle?   **1 July 1792**

To vulgar ears a black swan has the sound of a miracle:
but this arises merely from annexing the proverbial
name to the common swan, so emphatically distinguished by its constant snowy plumage, from which
it was never known to vary: but no one could ever
rationally be supposed to call in question the possible
existence of some distinct species of this numerous
genus, which, however nearly allied in point of size
and habit to the common swan, might yet be naturally
black. In fact such a species is now discovered. It is a
native of New Holland, and the neighbouring islands
. . . In general appearance it bears the most striking
resemblance to the common swan, and is remarkable
for all those gracefully-varying attitudes which so
eminently distinguish the European species.

New Holland   **George Shaw**[97]

Native name   **8 December 1798**

Those who amused themselves with shooting these
birds, found the parents an easy prey when attended
by their young; for nothing could induce them to quit
them, though fired at from every direction and
themselves frequently wounded, while their offspring
had life. The same affection has been noticed between
the male and female birds. The natives of New South
Wales name this bird Mulgo.

New South Wales   **Anon**[2]

This illustration from Shaw's mammoth *Naturalist's
Miscellany* (1789-1813)[97] was used later on decorative
chinaware.

Cape Barren goose

Cereopsis novaehollandiae

This bird has no close relatives, although fossils of a closely related bird have been found in New Zealand, and because of its unique features it is placed in a sub-family of its own. Cape Barren geese are among the least common of the world's waterfowl, and were at one time thought to be endangered, but they now number more than 6 000.

They are found on the southern coast of Western Australia and on the islands of Bass Strait, where their grazing habits have made them unpopular with farmers. They have been kept in captivity for many years and are represented in most zoological collections, but because of their pugnacity most zoos have only one pair. Unlike true geese, they need water only for drinking.

Cape Barren geese lay up to seven eggs, tending their young for six weeks until they join a flock of other juveniles. The adult is 700–1000 mm long.

First sighting   **12 December 1792**
. . . several allowed themselves to be taken by the hand; but the rest, apprized of the danger immediately flew away; this new species is somewhat smaller than our wild swan and of an ash-coloured grey, a little lighter on the belly. The bill is blackish, with a tumour of a sulphur-yellow at its base. The legs are slightly tinged with red.
Recherche Archipelago, W.A.   La Billardière[68]

Sighting in Bass Strait   **1798**
The goose approaches nearest to the species called *bernacle*; it feeds upon grass, and seldom takes to the water. I found this bird in considerable numbers on the smaller isles, but principally upon Preservation Island; its usual weight was from seven to ten pounds, and it formed our best repasts, but had become shy.
Furneaux Islands, Bass Strait   George Bass[6]

The species becoming scarce   **1860**
This bird is so exceedingly pugnacious, quarrelling with the poultry in the yard, as well as attacking pigs, dogs, or any other animals, that many persons, who have purchased the bird, have been glad to get rid of it. From not being bred in confinement, which could readily be done, they are becoming very rare, and are now seldom seen in Sydney. They breed almost every year in the Zoological Gardens in the Regent's Park [London], laying their eggs in March.
Sydney   George Bennett[10]

The 'Céréopse cendré' drawn by Huet for *Planches Enluminées de Buffon*.[109] By Conrad Temminck and Langier de Chartrouse, this was published serially between 1820 and 1838, to continue and complete the work that ceased fifty years earlier on the death of the French naturalist and philosopher, Compte de Buffon.

Jabiru

Xenorhynchus asiaticus

The jabiru or black-headed stork, Australia's only stork, is found across northern Australia and down the east coast, but is uncommon in the south. The same species extends from Australia through New Guinea, the Malay Archipelago and Burma to India.

Jabirus are found close to swamps, salt-water creeks and lagoons, and feed on fish, frogs, crabs, reptiles, rodents and carrion. The nest is formed from sticks and twigs topped with a thin layer of grass or rushes, in a tree so the sitting bird has a clear view of the surrounding area. Both the male and the female build the nest, incubate the eggs (2-4) and raise the young.

First scientific description — 4 December 1798
White Jabiru, with the head and neck green-black; the coverts, scapulars, and tail black; the bill black, the legs red.
New Holland   **George Shaw**[99]

Distribution of the jabiru — 1848
When the country was first colonized it was found as near to Sydney as Botany Bay, and even now is sometimes seen on the small islands in the mouth of the river Hunter; as we proceed eastward to Moreton Bay it becomes more common, and in the neighbourhood of the Clarence and MacLeay it may be almost daily seen: both Mr. Gilbert and Mr. McGillivray met with it at Port Essington, but did not procure specimens; the former also encountered it in the lagoons of the interior, while in company with Dr. Leichhardt . . . No bird is more shy in disposition or more difficult in approach, its feeding ground and resting-place being always in the most exposed situation, such as spits of land running out into the sea, large morasses, etc., where it can survey all around. Its food is said to be very varied, consisting of every kind of animal life inhabiting marshy situations, but more particularly fish and reptiles. Head and neck rich deep glossy green, changing into purple and violet at the occiput; greater wing-coverts both above and beneath, scapularies, lower part of the back and tail rich glossy green, tinged with a golden lustre; the remainder of the plumage pure white; bill black; irides dark hazel; legs fine red.
North coast of N.S.W. and Qld   **John Gould**[42]

Adventures of a captive jabiru — 1860
The Jabiru is an expensive bird to keep, consuming a pound and a half of meat daily, and being a very dainty feeder, the meat must be particularly fresh and good. When he was first placed in the yard where some poultry were kept, he stared at the fowls, and they ran away on his approach although he did not make the least attempt to molest them; and when striding round the yard, all the poultry fled before him, although it did not appear to be an intentional chase on his part. There happened to be a pugnacious, fussy little Bantam-cock in the yard, who would not permit the intrusion of any stranger, and on seeing the Jabiru he strutted up with expanded and fluttering wings and ruffled feathers . . . to frighten and drive him out of the yard. The Jabiru, with his keen bright eyes, regarded the little fluttering object with cool contempt, and walked about as before, the bantam followed. At last the Jabiru turned, and strode after the consequential little creature, as if to crush it under his feet; when the bantam, seeing matters take this serious turn, made off as fast as possible – like all little bullies – and did not venture to attack so formidable an opponent.
Sydney   **George Bennett**[10]

This small sketch of the head of a jabiru accompanied the first scientific description, made by Shaw, in 1788.[99] Shaw was director of the British Museum, London, until his death in 1813.

This hand-coloured illustration is from *Gatherings of a Naturalist in Australasia*[11] (published 1860) by George Bennett, a surgeon in early Sydney Town. Active in scientific circles, Dr Bennett became John Gould's agent in Sydney.

Wedge-tailed eagle

Aquila audax

First sighting   **6 August 1699**
There were but few Land-Fowls, we saw none but
Eagles, of the larger Sorts of Birds; but 5 or 6 Sorts of
small Birds.
Shark Bay, W.A.   **William Dampier**[30]

First capture   **March 1800**
Captain Waterhouse, in an excursion which he made
to the north arm of Broken Bay, wounded and
secured a bird, of a species never seen before in New
South Wales, at least by any of the colonists. It was a
large eagle, which gave a proof of his strength, by
driving his talons through a man's foot, while lying in
the bottom of the boat, with his legs tied together. It
stood about three feet in height, and during the ten
days that it lived was remarkable for refusing to be fed
by any but one particular person. Among the natives
it was an object of wonder and fear, as they could
never be prevailed upon to go near it. They asserted,
that it would carry off a middling-size kangooroo.
Captain Waterhouse hoped to have brought it to
England; but it was one morning found to have
divided the strands of a rope with which it was
fastened, and escaped.
Broken Bay, N.S.W.   **George Barrington**[5]

This wedge-tailed eagle crouching over its strange
prey was published in Barrington's history of New
South Wales, published in 1802 and like his sub-
sequent book 'Enriched with beautiful color'd
prints'.[5]

A further description   **7 November 1802**
The Mountain Eagle of New South Wales is a fine
majestic bird, which stands three feet high on the
ground.
The colour of its feathers is brown; the feet pale
yellow, and the talons, which it uses with the greatest
force, are black; the beak is of a yellowish horn colour;
and the crest, which is constituted by a few feathers,
has a yellow sandy appearance.
This bird is both an object of wonder and fear among
the natives, for it frequently takes up a kangaroo, a
dog, or a sheep; and probably they have little doubt
but that, if driven by hunger, and nothing else offered,
the Mountain Eagle would descend for the purpose of
carrying off a native.
New South Wales   **George Barrington**[5]

The largest bird of prey in Australia, with a wing-span of up to 2.5 metres, the wedge-tailed eagle eats mammals, reptiles and birds. Although its scientific name means the bold or audacious eagle, it is rare for one to take a lamb, despite farmers' fears.

This bird is found throughout Australia and Tasmania and builds a large nest of sticks, situated to give an uninterrupted view of the surroundings. Adults defend their territory against other eagles.

David Collins' *Account of the English Colony in New South Wales* (1798-1802)[28] included this illustration showing the eagle's strong beak and talons. As Judge Advocate, Collins had administered the oaths of office to Captain Phillip on the First Fleet's arrival in 1788.

Scrub fowl

Megapodius freycinet

Also known as a jungle hen, the scrub fowl is found in the coastal regions of the Northern Territory and in forests around the coast of Cape York and tropical eastern Australia.

The eggs are laid in large communal incubation mounds, and heat generated by the rotting vegetation which forms the mound incubates the eggs. The chicks kick their way out of the mound and from the time of their emergence are able to fly and fend for themselves.

Megapodius means 'big feet', a feature of great value to a bird which may spend up to six months tending an incubation mound. The mounds are useful food sources for the peoples of the tropics, as Lumholtz's account shows.

These two scrub fowls were painted for the French naturalist René Lesson, who published them in his *Zoologie* in 1838.[72]

Astounding mounds   1837

One discovery which was made through the medium of the natives, was that the large tumuli noticed by Captain King and others, and supposed to be raised by the inhabitants, are the works of a bird; some of them are thirty-feet long and about five feet high; they are always built near thick bushes in which they can take shelter, at the least alarm. The edifice is erected with the feet, which are remarkable both for size and strength, and a peculiar power of grasping; they are yellow while the body is brown. Nothing can be more curious than to see them hopping towards these piles on one foot, the other being filled with materials for building. Though much smaller in shape, in manner they much resemble moor-fowl. The use made of the mound is to contain eggs, which are deposited in layers, and are then hatched by the heat generated in part from decomposition. The instant that the shell bursts, the young bird comes forth strong and large, and runs without the slightest care being taken of it by the parent.

Northern Territory   **J. R. Stokes**[103]

How to cook the eggs   1890

November was just the time for the grauan [scrub fowl], which is found in great abundance in the lower part of the scrubs, but not higher up, where the cootjari [brush turkey] takes its place. The eggs are about four times the size of hens' eggs, and are prepared and eaten in the following original manner:

The blacks, having first made a hole on one side of the egg, place it on the hot ashes, and after a minute or two the contents begin to boil. Two objects are gained by making a hole in the egg – in the first place it does not break easily, and in the second place it can be eaten while lying boiling in the ashes. They dip into the egg the end of a cane that has been chewed so as to form a brush, and use this as a spoon.

Queensland   **Carl Lumholtz**[75]

This woodcut is from Lumholtz's book *Among Cannibals*[75], published in 1890; his zoological collection from Queensland went to the University of Christiana, Norway. Note the other fowl at work on the mound in the background.

Brush turkey

Alectura lathami

The brush turkey is found in the coastal forests of eastern Australia as far south as the Manning River, but being secretive is seldom seen.

Parkinson likened the bird to a 'Tetrao' which is a grouse or fowl-like bird. Dr Latham, who first described it scientifically, thought it was a vulture, because of the bare head and neck with wattles; later he noticed his error and renamed it *Alectura*, meaning 'cock's tail'. However, Swainson, in 1836, was still maintaining it was a 'New Holland vulture' in his *Classification of Birds*. The species belongs to the same group as barnyard fowls, as do grouse, so Parkinson's first impression was correct.

European scientists were able to see the interesting behaviour of this bird when several old birds constructed mounds and bred at Regent's Park Zoo in London.

First recorded sighting   4 July 1770
. . . a bird like a Tetrao, having wattles of a fine ultramarine colour, and whose beak and legs were black.
Endeavour River, Qld   Sydney Parkinson[84]

Brush turkey 'nests'   21 November 1822
Here the ranges were, for the most part, openly timbered, with the exception of the higher points, which were generally covered with vine-brush; in one of which we found the nests of the brush turkey . . . and observed the bird itself.
Murphy's Lake, Qld   Ludwig Leichhardt[71]

How to cook the eggs   1852
Many brush turkeys were shot by our sportsmen, and scarcely a day passed on which the natives did not procure for us some of their eggs. The mode in which these and other eggs are cooked by the blacks is to roll them up in two or three large leaves, and roast them in the ashes; the eggs burst, of course, but the leaves prevent the contents from escaping. Both bird and eggs are excellent eating; the latter, averaging three and a half inches in length, of a pure white colour, are deposited in low mounds of earth and leaves in the dense brushes in a similar manner to those of the megapodius, and are easily dug out with the hand. I have seen three or four taken out of one mound where they were arranged in a large circle, a foot and a half from the surface.
Cape York, Qld   John MacGillivray[77]

This illustration from John Gould's *Birds of Australia*[42] shows the bare red head and neck, with the bright yellow wattles – a feature that misled some early observers to describe the bird as a vulture. The length of the adult is about 700 mm.

Woodhen

Tricholimnas sylvestris

Shortly after the First Fleet arrived at Sydney Cove, the *Supply* was sent to Norfolk Island to obtain food. On the way, Lt Ball discovered Lord Howe Island, on 17 February 1788. The woodhen was found on the island, and as can be seen from the accounts the main interest in it was culinary.

Today it is one of the rarest birds in Australia, with only about thirty in existence, confined to the flat summit of Mt Gower on the island.

A member of the group of birds called rails, the woodhen is closely related to a flightless ground-bird of New Caledonia (now extinct), and more distantly to the weka of New Zealand.

First sighting   17 February 1788

We found no fresh water on the island, but it abounds with cabbage-palms, mangrove and manchineal trees, even up to the summits of the mountains. No vegetables were to be seen. On the shore there are plenty of ganets, a land-fowl, of a dusky brown colour, with a bill about four inches long, and feet like those of a chicken; these proved remarkably fat, and were very good food.

Lord Howe Island   Lt Henry Lidgbird Ball[90]

On the homeward journey   19 March 1788

The *Supply* in her return landed at the Island she made in going out & were very agreeably surpris'd to find great numbers of fine Turtle on the beach, & on the land amongst the trees great Nos. of Fowls very like a Guinea hen, & another Species of fowl not unlike the Landrail in England, & all so perfectly tame that you cd. frequently take hold of them with your hands but cd. at all times knock down as many as you thought proper wt. a short stick.

Lord Howe Island   Arthur Bowes[19]

This painting of the ground bird of Lord Howe Island is by John Hunter, a naturalist with the First Fleet. It is the best early illustration of the species. [61]

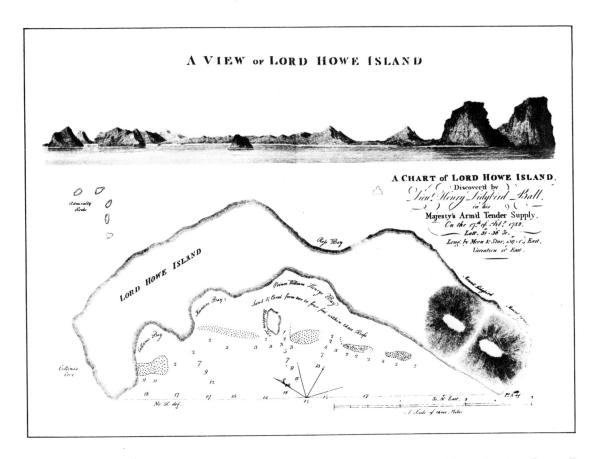

A VIEW OF LORD HOWE ISLAND

A CHART of LORD HOWE ISLAND.
Discover'd by
Lieut. Henry Lidgbird Ball.
in his
Majesty's Arm'd Tender Supply.
On the 17th of Feb.y 1788.
Latt. 31.36 S.
Long: by Moon & Star, 159.4 East.
Variation 11 East.

LORD HOWE ISLAND

Lt Henry Lidgbird Ball's 'view' and map of Lord Howe Island was published by Phillip.[90] 'The island is in the form of a crescent . . . Two points at first supposed to be separate islands, proved to be high mountains on the south–west end . . .'

Another account   16 May 1788

Partridges [woodhens] likewise in great plenty ran along the ground, very fat and exceedingly well tasted. Several of these I knocked down, and their legs being broken, I placed them near me as I sat under a tree. The pain they suffered caused them to make a doleful cry, which brought five or six dozen of the same kind to them, and by that means I was able to take nearly the whole of them. I might not otherwise have secured so many; for although they were by no means shy, yet they ran very fast when chased.
Lord Howe Island   **Thomas Gilbert[39]**

A further description   12 July 1788

However, it will be a Valuable Acquisition to the Colony at Port Jackson for it Abounds with Turtle Much superior to any I have ever seen, on the Shore we Caught several sorts of Birds, Particularly a Land fowl of a Dusky Brown About the size of a small Pullet, a bill 4 Inches Long & feet like a chicken.
Lord Howe Island   **David Blackburn[15]**

Becoming rare   August 1887

Soon to become extinct on Lord Howe, unless protected, is the Wood-Hen . . . a curious and stupid bird. At the present time its range is confined to the extreme southern end of the island, in Erskine Valley, and the ground around the sea-girt base of Mount Gower. It is even now rare and difficult to obtain, and would be impossible of capture were it not for the fact that its curiosity overcomes its shyness. Its gradual extinction is probably due to the ravages committed by the wild domestic cats. During a journey to Mount Gower, primarily to procure specimens, only one individual was seen, and during the whole of our residence there those well acquainted with their haunts could obtain but four others. [The Wood-Hen] can be attracted within gunshot by any continuous and varied noise, such as knocking two stones together, striking against a tree, occasional whistling, and other peculiar but discordant noises.
Lord Howe Island   **Robert Etheridge[36a]**

By 1928, when G. M. Mathews' book was published[81], the woodhen had become scarce as a consequence of predation by people, wild pigs and dogs. The posture of Mathews' bird is more accurate than Hunter's.

Bustard

Ardeotis australis

Early contact with the bustard shows that the settlers' main interest in it was as a food supply. It is common now only on the Cape York Peninsula, the Barkly Tableland, in the Kimberleys and on the Nullarbor Plain – though it could once be seen almost anywhere in open country on the mainland.

A large bird – more than a metre long – the bustard is easily shot because it will freeze when disturbed and even if pursued will only walk away slowly. It eats grasses, seeds, fruits, grasshoppers, crickets, small mammals and reptiles; and the female lays only one or two eggs.

First sighting and shooting   23 May 1770
All or most of the same sort of land and water fowl as we saw at Botany Bay we saw here, besides these Black & white Ducks, and Bustards such as we have in England one of which we killd that weigh'd 17½ pounds which occasioned my giving this place the name of *Bustard Bay*.
Bustard Bay, Qld   James Cook[7]

First eating   24 May 1770
At Dinner we eat the Bustard we had shot yesterday, it turnd out an excellent bird, far the best we all agreed that we have eat since we left England, and as it weighd 15 pounds our Dinner was not only good but plentyfull.
Bustard Bay, Qld   Joseph Banks[8]

Another good meal   1844
This fine and erectly walking bird is also common over the whole of the interior, migrating from the north in September and October. Several flights of these birds were seen by us thus migrating southwards in August, passing over our heads at a considerable elevation, as if they intended to be long on the wing. I have known this *Otis* weigh 28 lbs. Its flesh is dark and varied in shade. The flavour is game and the meat is tender.
Central Australia   Charles Sturt[104]

A wary bird   1846
One day I killed a bustard . . . weighing 21½ pounds; the goodness of its flesh was duly appreciated by my messmates. Several small flocks of this noblest of the Australian game-birds were seen; but, from their frequenting the open country, and being very wary, it is only by stratagem or accident that they can be approached within gunshot.
Queensland   John MacGillivray[77]

The first major works published on the birds and mammals of Australia were by the Englishman John Gould (1804–81). His artists included his wife (who died in 1841, after their journey to Australia), H. C. Richter, and Edward Lear.[42]

Brolga

Grus rubicundus

The brolga, or native companion, is the only true crane found only in Australia. A stately bird, it was once widely distributed over most of the northern and eastern Australian swamplands, but it is now uncommon in the south-east; it can still be seen in large numbers, however, on Townsville Common in north Queensland.

The brolga is at its most elegant when dancing – stepping back and forth on its long legs, shaking its half-open wings. The dancing is generally a courtship display, though it sometimes takes place outside the breeding season. The bird feeds on insects, frogs and sedge tubers and is generally found near water, where the platform nest is built.

First sighting　　　　**August 1770**
Birds there were Several Species of . . . but tho we saw many thousands of them were so shy that we never got one of them; as were the Cranes also of which we saw several very Large and some beautiful species.
North Queensland coast　　　**Joseph Banks[8]**

Native companion　　　　**1844**
This large sized Crane is common near the waters of the interior, but he is a wary bird, and seldom lets the fowler within shot. When seen in companies they often stand in a row, as they fly in a line like wild fowl. Their general plumage is slate colour, but they have a red ceres or skin on the head. One of these birds was tame in the Government domain at Parramatta in 1829.
Central Australia　　　**Charles Sturt[104]**

An early dissection　　**24 October 1844**
On dissecting the *Grus* . . . I found the convolution of the Trachae although a female bird – the keel of the sternum is hollowed out as in some of the European Swans – the trachae reaching to near three inches of its length before it turns back to enter the lungs. The stomach was exceedingly thick and muscular and contained large pebbles in great proportions, seeds of the swamp grass, and Coleoptera, and what appeared to be vegetable matter generally. The flesh of this bird we had for breakfast, and found it not only excellent but sufficient.
Queensland　　　**John Gilbert[39]**

This illustration of a brolga is from John Gould's *Birds of Australia* (1848)[42]. The scientific name for the species, *rubicundus*, refers to its red head.

Sulphur-crested cockatoo

Cacatua galerita

This well-known parrot is found from south-eastern Australia northwards and across the 'top end' to north-western Australia. Other races of this species occur in New Guinea.

In open country, where the birds often congregate in large flocks, they have a sentinel warning system: several birds, perched in nearby trees, stand guard while the rest of the flock feed on the ground; at the sign of anyone approaching, the sentinel birds screech loudly and the flock then takes to the air.

The bird's natural diet includes seeds, berries, fruits, nuts, flowers, leaf buds, roots and insects and their larvae. This should be remembered when keeping one as a pet; raw sweet corn cobs, cuttlefish bone, apple, raw peanuts, fruit-tree or eucalyptus prunings, spinach and lettuce should be added to the basic 'cockies' seed mixture. (Length: 490 mm)

An early description   1790

The bird seems liable to great variation both as to size and colour; the white in some being of a much purer appearance than in others, and the yellow on the crest and tail more predominant. All the varieties yet known agree in having the beak and legs blackish. The individual specimen here figured seemed of a somewhat slenderer form than usual. The colour not a pure white, but slightly tinged on the upper parts, and particularly on the neck and shoulders, with dusky. The feathers on the front white, but the long lanceolate feathers below them, which form the crest, of a pale jonquil-yellow. The tail white above, and pale yellow beneath; as are also the wings.

New South Wales   **John White**[115]

The bird's habits   1799

We saw this beautiful Cockatoo alive. It expresses joy, by shaking its head briskly several times upwards and downwards, marking a slight cracking with its bill, and displaying its elegant crest. It returns caresses; touches the face with its tongue, and seems to lick it. The kisses are soft and gentle. When one hand is laid flat under its body, and the other rests on its back, or only touches its bill, it presses firmly, claps its wings, and, with its bill half open, it blows and pants, and seems to feel the most intoxicating delights. It repeats this as often as we please. It is also very fond of being

This picture from the *Naturalist's Pocket Magazine*[2] of 1799 is a direct steal from Shaw's *Naturalist's Miscellany*.[97] Nodder, the artist of Shaw's book, possibly illustrated both works.

scratched, and holds its head, and raises its wing, to be stroked. It often whets its bill, by gnawing and breaking bits of wood. It cannot bear the confinement of the cage, but it never roves out of its master's sight. It answers his call, and retires when he commands: in which last case, it discovers anxiety, often looking back for the sign of invitation. It is exceedingly neat; and all its motions are graceful, delicate and pleasing. It feeds on fruits, pulse, all the farinaceous grains, pastry, eggs, milk, and whatever is sweet, if not too sugary.

New South Wales   **Buffon**[2]

Cockatoo lookout   1844

This Cockatoo . . . the most mischievous of Australian birds, and . . . plays sad havoc amongst the wheat when ripe . . . They are in immense flocks, and when in mischief always have sentinels at some prominent point to prevent their being taken by surprise, and signify the approach of a foe by a loud scream.

Central Australia   **Charles Sturt**[104]

The world's oldest bird   early 1900s

The world's oldest bird was Cocky Bennett, of Tom Ugly's Point, a Sydney riverside suburb. He was believed to be about 110 years old when he died. There was no doubt at all that he was at least a centenarian. Near the end of his life he was almost naked, with only two or three forlorn feathers draping his shrivelled brown body, the sad remains of a once impressive yellow crest. He may have lost his feathers, but right to the end he remained very talkative and bossy.

Tom Ugly's Point, N.S.W.   **Anon[3]**

Gould's *Birds of Australia*[42] illustrates two sulphur-crested cockatoos, one half-hidden in the nesting hole; the nest is usually built in a eucalyptus, near water, and two to three eggs are laid.

Red-tailed black cockatoo

Calyptorhynchus magnificus

The red-tailed black cockatoo is found all over the continent, with the exception of the deserts of South Australia and Western Australia. This parrot is primarily a seed-eater but it also sometimes feeds on insect larvae, which it extracts from tree branches. One or two eggs are laid in the nest built in a tree hollow.

The bird was originally observed and collected on the *Endeavour* expedition of Captain Cook in 1770, and taken to England by Joseph Banks.

When George Shaw described the species, he noticed considerable variation in plumage in different specimens, and correctly guessed that these were differences between males and females. (Length: 500-610 mm)

First sighting   4 July 1770

Of birds we found . . . large black cocatoes, with scarlet and orange-coloured feathers on their tails, and some white spots between the beak and the ear, as well as one on each wing.

Endeavour River, Qld   Sydney Parkinson[84]

Sydney Parkinson, Joseph Banks' artist on the *Endeavour* made this sketch in 1770.[85] It was the first European drawing of a black cockatoo.

Another Endeavour sighting　　　**August 1770**

The Land Birds were crows, very like if not quite the same as our English ones, Parrots and Paraquets most beautifull, white and black Cocatoes, Pidgeons, beautifull Doves, Bustards, and many others which did not at all resemble those of Europe. Most of these were extremely shy so that it was with difficulty that we shot any of them; a Crow in England tho in general sufficiently wary is I must say a fool to a New Holland crow and the same may be said of almost if not all the Birds in the countrey.

North Queensland coast　　　**Joseph Banks**[8]

First scientific description　　　**1787**

Size of the *Red and Blue Maccaw*: length twenty-two inches. Bill very large, of a horn-colour, with a black tip: general colour of the plumage black: the feathers of the head pretty long, but in a quiescent state lie flat on the head; on each, just at the tip, is a spot of pale buff-colour: the wing coverts are also marked in the same manner near the tips: the feathers of the upper part of the breast and vent are margined with buff; the lower part of the breast and the belly barred with darker and lighter buff-colour: the tail is pretty long, and a little rounded at the end: the two middle feathers are black; the others the same at the base and ends; the middle of them, for about one third, of a fine deep crimson, inclining to orange, crossed with five or six bars of black, about one third of an inch in breadth, and somewhat irregular, especially the outer ones, in which the bars are broken and mottled: legs black. Inhabits New Holland. In the collection of Sir Joseph Banks who brought it with him from thence into *England* on his return from his voyage round the world. It most certainly differs from the Ceylonese Black Cockatoo; but is probably the same with that mentioned by Mr. Parkinson, in his voyage.

New Holland　　　**John Latham**[69]

This bird is a female, as the tail feathers are barred; males have a single wide scarlet band. The picture was published with John Latham's description of the species.[69] Latham, a London doctor and contemporary of Joseph Banks, was the first to describe and name many Australian species.

A further description　　　**1 December 1790**

New Holland, which may not unjustly lay claim to the more dignified title of the Southern Continent, may be considered as a kind of new world to the Naturalist; and has already afforded several animals unknown to every other part of the globe. To no other genus, however, have such large accessions of new species been added as to that of Psittacus; of which some of the most superb kinds appear to be natives of New Holland, and some of the Southern Islands. Of these newly-discovered birds, one of the most august in its appearance is that represented on the plate annexed. In size it is equal to the great Maccaws already so well known to Naturalists; but it belongs to a different section in the genus, and instead of being furnished with a lanceolate tail, as in those birds, it has that part even at the end, or consisting of feathers of nearly equal length . . . The whole bird is of so superior a magnificence as justly to be regarded as one of the finest of its tribe. It is subject to some variation, and in some specimens the scarlet on the tail, instead of being barred with black, forms one large transverse band in the middle.

New Holland　　　**George Shaw**[99]

Rainbow lorikeet

Trichoglossus haematodus

The rainbow lorikeet is the Australian representative of a species called coconut lorikeet; it is found on the eastern coast, from Torres Strait south to Tasmania, and across to Kangaroo Island and the Eyre Peninsula.

The brush-tipped tongue, which all lorikeets possess, is for collecting nectar and pollen from flowers. The bird is easily tamed, and a suitable substitute diet is a mixture of raw sugar, infant cereal, wholemeal bread and water supplemented with fruits. (Length: 320 mm)

First capture   **May 1770**

Tupaia caught a rainbow lorikeet and took it aboard the *Endeavour* as a pet. It survived, outliving its owner who died on 26 December 1770, and was painted from life in England by Peter Brown.
Botany Bay   **[Eds]**

First description   **3 November 1774**

Length, fifteen inches.
Bill,   of a reddish colour.
Head,   of a rich dark blue, beautifully mixed with small feathers, of a light blue.
Orbits, black.
Neck,   towards the throat of a yellowish green; the hind part, green.
Breast, red mixed with yellow.
Belly,   of a fine blue.
Thighs, green and yellow.
Back,   and *Wings*, green; the *Primaries* dusky, barred with yellow.
Tail,   cuneiform, Middle feathers, green; the rest, green; on their exterior sides, of a bright yellow.
Legs,   dusky.
Place,   New South Wales, in New Holland; very numerous in Botany Bay.
           This bird was first brought over by Joseph Banks, Esq.
Botany Bay   **Peter Brown**[21]

The first eastern Australian bird to reach England alive, painted by Peter Brown, artist and zoologist attached to Marmaduke Tunstall's museum (later the Newcastle Museum).[21]

Sighting by First Fleeters   **April 1788**

We likewise saw several Blue-bellied Parrots. This is a very beautiful bird; and is a very common species in various parts of *New Holland*, and in great plenty both at *Botany Bay* and *Port Jackson*.
Port Jackson   **John White**[115]

Parrots with brush-tongues   **1860**

In New South Wales there is a very numerous family of Honey-eaters or Meliphagidae (elegant and gaily-coloured birds), and Parrakeets of brilliant plumage and rapid flight, differing from the slow, jerking action of other Parrots, – forming the genus *Trichoglossus*, or Honey-eating Parrakeets; these have a feathered brush-like tongue, but no gizzard; their crops are found filled with honey, from the flowers of the *Eucalyptus* or gum-tree, on which they feed.
Sydney   **George Bennett**[11]

Budgerigar

Melopsittacus undulatus

This small bird is distributed over most of inland Australia and is probably the most numerous parrot species in the country.

The budgerigar feeds on grass seeds, and in good seasons huge flocks gather, moving from place to place according to the availability of food and water. (Length: 180 mm)

Shaw's first description   1805

The highly elegant species of Parrakeet represented on the present plate in its natural size, is an inhabitant of New Holland, and seems to have been hitherto undescribed. The upper parts of the bird, from the bill to the rump, are of a pale yellowish green, beautifully crossed by numerous linear brown undulations, which become gradually larger as they approach the back and shoulders: the wing-feathers are brown, with pale olive-yellow edges: the under parts of the bird, together with the rump, are of an elegant pale green; the throat pale yellow, mottled on each side with a few small deep blue scattered spots, accompanied by small black crescents: the tail is of a cuneated form, and of a deep-blue colour, with a bright yellow bar running obliquely across all the feathers except the two middle ones, which considerably exceed the rest in length: the bill and legs are brown.

New Holland   **George Shaw**[97]

First budgerigars captured   1838

On arriving at Brezi, to the north of Liverpool Plains, in the beginning of December, I found myself surrounded by numbers, breeding in all the hollow spouts of the large *Eucalypti* bordering the Mokai; and on crossing the plains between that river and the Peel, in the direction of the Turi Mountain, I saw them in flocks of many hundreds feeding upon the grass-seeds that were there abundant. So numerous were they, that I determined to encamp on the spot, in order to observe their habits and procure specimens.

Liverpool Plains, N.S.W.   **John Gould**[42]

Sturt's observations   1838

Called "Bidgerigung" by the natives. This beautiful little Euphema visits South Australia about the end of August or the beginning of September, and remains until some time after the breeding season. It is perhaps the most numerous of the summer birds. I remember, in 1838, being at the head of St. Vincent's Gulf, early in September, and seeing flights of these birds, and *Nymphicus Novae-Holl.* following each other in numbers of from 50 to 100 along the coast line, like starlings following a line of coast. They came directly from the north, and all kept the same straight line, or in each other's wake. Both birds subsequently disperse over the province. The plumage of this bird is a bright yellow, scolloped black, and three or four beautiful deep blue spots over each side the cheek.

St Vincent's Gulf, S.A.   **Charles Sturt**[104]

Budgerigars as cagebirds   1860

A very delicate and beautiful little Parrot, having a wide range over New South Wales, and also now well known in England, from numerous specimens having been sent thither, is the Canary or Zebra Parrot, Warbling Grass Parrakeet or Love-bird of the colonists; it is the Budgeree-gar of the aborigines – *Budgeree* signifying handsome or good. I recollect well when my friend Mr. Gould brought these birds (the first procured alive) to my house, during his visit to Sydney in 1839. They were delicate, and two died from exposure near a window; but he succeeded in bringing the remainder safe to England.

New South Wales   **George Bennett**[11]

This illustration, the first published picture of a budgerigar, is from Shaw's *Naturalist's Miscellany* of 1805.[97]

Gould's illustration shows a pair of budgerigars
feeding on grass seeds.[42] Birds have no teeth, and food
bitten off by the beak in rather large particles is
ground down by grit in a muscular gizzard.

Eastern rosella

Platycercus eximius

The eastern rosella was discovered and examples of the species were sent back to England soon after the First Fleet arrived. The common name of the group of broad-tailed parrots, the rosellas, is a corruption of 'Rosehiller', the name the early colonists gave the species, since it was at Rose Hill that the bird was first seen. This rosella was named *Platycercus eximius* by George Shaw – *eximius* meaning excellent.

It is a familiar bird in open timbered country on the outskirts of cities and towns throughout its south-east Australian range. It feeds on berries, thistles and seeds. (Length: 300 mm)

The 'Nonpareil Parrot' described   1792

Long-tailed variegated parrot, with head throat breast and vent crimson, back black undulated with yellow-green, blue wings and tail. The two middle tail-feathers are green. This bird is a species hitherto undescribed; having been very lately brought from New Holland. To particularize the richness of its robe would be unnecessary; the figure accurately shewing all its variegations of colour. In size and general form it is strongly allied to the Pennantian Parrot [crimson rosella].

New Holland   **George Shaw**[97]

An 'admired animal'   1794

It is in these savage regions however, that Nature seems to have poured forth many of her most highly ornamented products with unusual liberality: where, in particular, she appears to have stationed birds, superior perhaps in elegance to those of most other climes; and where their beauties can only be contemplated by the eyes of barbarians. Amidst the number of these admired animals, the species represented on the annexed plate may justly claim a distinguished place.

New Holland   **George Shaw**[98]

Taken at Kissing Point   7 August 1804

A Parrot of a species perfectly distinct from any hitherto found, was lately taken at Kissing Point, and is now in the possession of the Judge Advocate. Its size differs little from that of the Lowry, but the feather is by no means the same: those of the neck and breast are of a rich scarlet, with the head, wings and tail of a clear straw colour.

Sydney   **Sydney Gazette**[106]

Talking parrots   1834

Parrots are, perhaps, of all the feathered tribe, the most numerous in the colony; and different species are lauded for speaking, whistling and other accomplishments. No one can walk the streets of Sydney or any of the villages of the colony, or enter an inn or dwelling-house, whithout seeing this class of birds hung about in cages, and having his ears assailed by the screeching, babbling and whistling noises which issue from their vocal organs: it is the sweet music of the colony, and "pretty polly", "sweet polly", are tender sounds which issue from the exterior as well as interior of every dwelling.

New South Wales   **George Bennett**[10]

The first European description of the bird, by George Shaw[97], was accompanied by this illustration. It is an unnatural stance; the bird does not elevate its tail. And like many early illustrations, it shows European vegetation – rather than Australian.

Crimson rosella

Platycercus elegans

This parrot is common throughout south-eastern Australia, with an isolated population in north-eastern Queensland. Rosellas were taken to England with the *Endeavour* in 1770, and were apparently given by Joseph Banks – with other animals from the expedition – to a leading zoologist, Thomas Pennant. John Latham described the bird as *Psittacus pennanti*, and its common name as an aviary bird in England is still Pennant's parakeet. However, Gmelin named the bird in his *Systema Naturae* before Latham, and so his scientific name, *Platycercus elegans* (the elegant broadtail), has precedence and is the one used today.

The crimson rosella has green feathers in the juvenile stage, and early scientists thought that the green birds were females. In fact, both sexes are coloured identically.

The bird eats seeds, fruits, blossoms, insects and larvae, and nests in hollow logs or branches.

An early scientific description   1787

Length fifteen inches. Head, lower part of the back, and all the under parts of the body, scarlet: chin of rich blue: upper part of the back, and scapulars, deep brown, or black, edged with scarlet: lesser coverts pale blueish green: ends and interior sides of the quills dusky, marked on the inner webs with a single white spot: sides deep blue: tail very long, the middle feathers dusky; the exterior and upper part of the interior sides blue; the other parts of fine green; tips of the exterior feathers white.

New South Wales   John Latham[69]

Rediscovered by the First Fleet   6 August 1789

This beautiful bird is not unfrequent about *Port Jackson*, and seems to correspond greatly with the Pennantian Parrot, described by Mr. Latham in the supplement to his General Synopsis of Birds, p. 61. differing in so few particulars, as to make us suppose it to differ only in sex from the species.

Botany Bay   Arthur Phillip[90]

Peach yield affected by parrots   7 April 1821

By the immense numbers of parroquets, which frequented the Settlement during last winter and till spring, the blossom buds of the peach trees were generally plucked off; and the consequence is, little or no fruit of this kind has been seen this year.

Sydney   Sydney Gazette[108]

The First Fleet ships that returned from Port Jackson to England carried many new plants and animals for study. The illustrators worked from flat skins and often had little idea of the true stance of the animal. Phillip's illustration of 1789 was drawn from a study skin.[90]

Crimson rosellas in Gould's *Birds of Australia*[42] in a lifelike pose and realistically depicted in a casuarina tree: in front is an adult, with an immature green bird behind.

Paradise parrot

Psephotus pulcherrimus

This parrot was not discovered until 1844 when John Gilbert, one of Gould's collectors, found it just before joining Ludwig Leichhardt on the ill-fated expedition to Port Essington. The bird was captured on the Darling Downs and sent to John Gould, whose artists illustrated it for *Birds of Australia*. Gilbert's diaries are in the Mitchell Library, but the manuscript is so faint that his 'first sighting' account is illegible.

Gould's scientific name for this species, *pulcherrimus*, means 'prettiest'. (Length 300 mm)

First description **20 Broad Street, Golden Square 11 January 1845**

To: R. Taylor
Zoological Society of London

Dear Sir,

My collector, Mr. Gilbert, has lately sent me the description of a new *Platycercus* discovered on the Darling Downs at the back of Moreton Bay, on the east coast of Australia, and which he states far surpasses in beauty every other species of the genus yet discovered. I have therefore thought it of sufficient importance to the ornithologist to send you a copy for insertion in the 'Annals of Natural History' . . . In habits it is a truly grass-feeding parrakeet. For this beautiful species I propose the name of *Platycercus pulcherrimus*.

John Gould[42]

Purchased by the London Zoo **10 April 1866**

Mr. P. L. Sclater, Secretary to the Society, called the attention of the Meeting to a pair of one of the most beautiful of all the Australian Parrakeets recently added to the Aviaries for the first time. These birds had been brought over by Mr. N. Timmermann, Steward of the ship 'Nineveh', on the 20th of March, and purchased of him for the collection.

Zoological Society of London[95]

Parrot behaviour **1889**

In this region I shot two specimens of the beautiful parrot *Platycercus pulcherrimus* under the following remarkable circumstances. An hour before sunset I left the camp with my gun, and soon caught sight of a pair of these parrots, a male and a female, that were walking near an ant-hill eating grass-seed. After I had shot the male, the female flew up into a neighbouring tree. I did not at once go to pick up the dead bird – the fine scarlet feathers of the lower part of its belly, which shone in the rays of the setting sun, could easily be seen in the distance. Soon after the female came flying down to her dead mate. With her beak she repeatedly lifted the dead head up from the ground, walked to and fro over the body, as if she would bring it to life again; then she flew away, but immediately returned with some dry straws of grass in her beak, and laid them before the dead bird, evidently for the purpose of getting him to eat the seed. As this too was in vain, she began again to raise her mate's head and to trample on his body, and finally flew away to a tree just as darkness was coming on. I approached the tree, and a shot put an end to the faithful animal's sorrow.

Nogoa River, Qld **Carl Lumholtz[76]**

The illustration of the paradise parrot in Gould's great work[42] is especially valuable. The last authenticated sighting was in 1922 in the Burnett River region of east Queensland. Hearsay reports since then have been recorded, but there is a possibility that the parrot is extinct.

Superb lyrebird

Menura novaehollandiae

There are two species of lyrebird – the superb lyrebird and Albert's lyrebird – but only the former is illustrated here. Lyrebirds are the largest perching birds in the world: up to 970 mm long, including the tail. As well as having a range of their own songs, they mimic the calls of other species, sometimes even imitating the sounds of power saws and woodchopping. Males may keep up to ten display grounds in their defended territories, on which they dance and sing during the mating season.

Because of their secretive behaviour, and the heavily wooded gullies in which they live, lyrebirds are unlikely to be seen except by the most careful observer. They feed on insects and small animals, which they scratch from the forest floor and rotten logs with their large feet.

First European sighting   November 1797

An ex-convict who lived with Aboriginals after his term expired in 1792, said that there was in the bush near Sydney, "a bird of the pheasant species".

Near Sydney   John Wilson[5]

First recorded sighting   24 January 1798

We saw nothing strange except a few rock kangaroos with long black brush tails, and two pheasants which we could not get a shot at.

Nepean, N.S.W.   John Price[92]

First capture   26 January 1798

Here I shot a bird about the size of a Pheasant, but the tail of it very much resembles a Peacock, with large long feathers which are white, orange, and lead colour, and black at the ends; its body betwixt a brown and green, brown under his neck and black upon his head. Black legs and very long claws.

Near Bargo, N.S.W.   John Price[92]

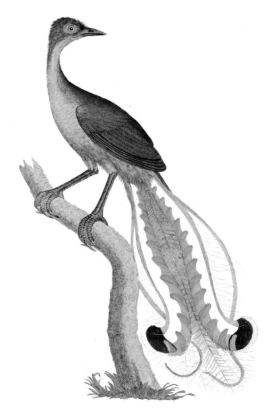

From New South Wales, Thomas Davies sent his 1799 painting, a skin and a written description to Joseph Banks, who presented them to the Linnean Society on 4 November 1800 as *Menura superba*. Before they were published in the Society's journal (in 1802)[31], John Latham had included the new bird in his supplement to *Index Ornithologicus* as *Menura novaehollandiae*.

The mimicry of the lyrebird · March 1798

They frequent retired and inaccessible parts of the interior; have been seen to run remarkably fast, but their tails are so cumberous that they cannot fly in a direct line. They sing for two hours in the morning, beginning from the time when they quit the valley, until they attain the summit of the hill; where they scrape together a small hillock, on which they stand, with their tail spread over them, imitating successively the note of every bird known in the country. They then return to the valley.

South-west of Sydney   **David Collins**[28]

An early scientific description
4 November 1800

The total length of this singular bird from the point of the bill to the end of the broad tail feathers is 43 inches; 25 of which are in the tail alone. The bill rather exceeds an inch in length, is strong, formed much like that of a peacock . . . From these birds being found in the hilly parts of the country, they are called by the inhabitants the Mountain Pheasant. With respect to their food or manners I have not as yet obtained any particular account. In my specimens, there is a nakedness round the eyes, but whether this is from the feathers having fallen off I know not. I rather think otherwise, and that it may be brightly coloured as in many other birds.

Blackheath, N.S.W.   **Thomas Davies**[31]

Barrington's illustration[5] shows artistic licence – the tail is about twice the size it is in a living specimen. Barrington himself was a gentleman pickpocket transported to New South Wales – there to be described variously as a religious convert, chief constable and lunatic.

Superb blue wren

Malurus cyaneus

The superb blue wren belongs to a group found only in Australia: the fairy wrens, or wren-warblers. The word 'wren' is misleading, because the true European wren belongs to a different family of birds and is not found in Australia; the early settlers nostalgically named many birds after those they knew at home.

The dull-coloured female lays three to four eggs in a nest constructed in a bushy shrub. The babies remain with the parents for a year – which gave rise to the notion that the male was polygamous. In very good seasons the female may nest again, and while she is incubating the eggs the male takes over the rearing of the first group. The superb blue wren is found in the south-east of the mainland and in Tasmania. The adult is 160 mm long.

First observed   **29 January 1777**

The length five inches. The bill straight, subulated, & of a jet black. The forehead, neck behind, the temple of a beautiful bright azure. The back, space before the eyes, & tail coverts, of a jet or velvet black. The throat, breast and tail of a deep blue or violet colour. The legs blackish. The tail as long as the body with the feathers of an equal length.

Adventure Bay, Tas.   **William Anderson[1]**

The female is brown,
the male blue   **17 June 1789**

The female is now discovered to be entirely destitute of all the fine blue colours, both pale and dark, by which the male is adorned, except that there is a very narrow circle of azure round each eye, apparently on the skin only: all the upper feathers consist of shades of brown, and the whole throat and belly is pure white. Except from the shape and size, this bird would not be suspected at first sight to belong to the same species as the male: the epithet of superb applies very ill to the female.

New South Wales   **Arthur Phillip[90]**

Soft and elegant blue   **1 November 1789**

The beautiful species of Motacilla here figured, is a native of that part of New Holland called Van Diemen's Land; and is one of the new species of birds, which have been discovered during the voyages to those parts. It varies a little in colour, some specimens having more of the blue on the head than others; the belly also in some specimens is of a more dusky tinge than in others. The head is of the deepest velvet-black, and the feathers on the whole bird have an unusual share of softness and elegance.

Van Diemen's Land   **George Shaw[97]**

The earliest illustration of a superb blue wren.[36] It appeared in 1777 and was drawn by W. W. Ellis, assistant-surgeon on Cook's third voyage, at Adventure Bay in Tasmania. The tail shape and position is unfortunately more appropriate to a flycatcher.

More accurate than the first European illustration, this engraving of a female wren was published in 1789 in Phillip's *Voyage to Botany Bay*.[90] To the illustration the engraver has added a boronia-like plant – perhaps working from leaves and flowers pressed in New Holland for English botanists.

John White's honeyeater eyeing two insects. In 1790,
when this picture was published[115], the yellow-faced
honeyeater was thought to eat only insects and was
described as a 'flycatcher'.

Yellow-faced honeyeater

Lichenostomus chrysops

The yellow-faced honeyeater is one of the commonest honeyeaters, extending from the Atherton Tableland in Queensland to Mount Lofty and the Flinders Ranges in South Australia. Not surprisingly, it was one of the first honeyeaters encountered by the early settlers, and White's illustration is particularly accurate; its musical call was often noted in early accounts. Its food consists of insects from the outer foliage of trees, and nectar from the flowers of various grevilleas, banksias and eucalypts.

An early sighting **17 May 1788**

We this day caught a Yellow-eared Flycatcher (see annexed plate). This bird is a native of New Holland, the size of a martin, and nearly seven inches in length; the bill is broad at the bottom, and of a pale colour; the legs dusky; the plumage is mostly brown, mottled with paler brown; the edges of the wing feathers yellowish; the under part of the body white, inclining to dusky about the chin and throat; the tail is pretty long, and, when spread, seems hollowed out at the tip; beneath the eye, on each side, in an irregular streak, growing wider, and finishing on the ears, of a yellow or gold colour.

Port Jackson **John White**[115]

Song of the honeyeater **24 October 1844**

On observing the two species of *Ptilotis* [honeyeater] I have killed, I remarked during a ramble this afternoon, that *P. chrysops* [yellow-faced honeyeater], posses a very loud and at times rather pleasing note, on the whole very much resembling *Glyciphila ocularis* [brown honeyeater], while *P. fusca* [fuscous honeyeater] has only a succession of tweet tweet like notes.

Queensland **John Gilbert**[39]

Nest **1848**

A nest found near the Liverpool range in October was very neatly constructed, rather small in size, round, and so thin that I could see through it; it was suspended to the fine twigs of a *Casuarina* at some height from the ground, while another suspended to the lower branches of a sapling gum was within reach of the hand. They were outwardly composed of the inner bark of trees, moss, etc., lined with fine vegetable fibres and grasses. The eggs, which are two and sometimes three in number, are of a lengthened form, and of a deep reddish buff, strongly marked at the larger end with deep chestnut-red and purplish grey; the remainder of the surface ornamented with large spots and blotches of the same colour, somewhat thinly dispersed; their medium length is ten lines and a half by seven lines in breadth.

Liverpool Range, N.S.W. **John Gould**[42]

A pair of honeyeaters, perched on a spray of native sarsparilla (from Gould's *Birds of Australia*[42]).

New Holland honeyeater

Phylidonyris novaehollandiae

The honeyeaters are a family of birds with brush-tipped tongues which enable them to feed on the nectar and pollen of flowers. None is entirely nectivorous; most eat insects associated with the flowers, and a few are decidedly insectivorous.

The New Holland honeyeater is a common bird and was first recorded by John Latham, from the diary of W. W. Ellis, assistant-surgeon to Captain Cook's third voyage. It was first called the New Holland creeper because of its ability to hang upside-down on fine stems when feeding on flowers. It is a bird of the heathlands and open coastal forests of the south-eastern and south-western mainland and Tasmania, where it feeds on banksia, eucalyptus and grevillea blossoms.

The usual colouring of the New Holland honeyeater, showing underparts white streaked with black, from the journal of John White, chief surgeon of the First Fleet. [115]

First popular description　　　　　**July 1788**

We discovered the *New Holland Creeper*; (see plate annexed). The general colour of the bird is black, spotted in various parts with white: the bill is dusky, growing paler towards the tip. The neck, breast, belly, and sides are more or less streaked with white; over the eye is also a white streak, and the sides of the neck and beginning of the back have likewise some streaks of the same. The quills and tail feathers are marked with yellow on the outer margins; the last are rounded in shape, and two or three of the outer feathers spotted within, at the tip, with white; legs dusky; is about the size of a *nightingale*, and measures seven inches in length.

Port Jackson　　　　　**John White**[115]

Nests at Government House　　　　　**1848**

The *Meliphaga Novae-Hollandiae* [New Holland honeyeater] is one of the most abundant and familiar birds inhabiting the colonies of New South Wales, Van Diemen's Land, and South Australia: all the gardens of the settlers are visited by it, and among their shrubs and flowering plants it annually breeds. It has a loud, shrill, liquid, although monotonous note. Its food, which consists of the pollen and juices of flowers, is procured while clinging and creeping among them in every variety of position: it also feeds on fruits and insects. It usually rears two or three broods during the course of the season, which lasts from August to January: the nest is very easily found, being placed, in the forest, in any low open bush, and in the gardens among the shrubs and flowers: one of the nests in my collection was taken from a row of peas in the kitchen-garden of the Government House at Sydney. It is usually placed at about eighteen inches or two feet from the ground, and is a somewhat compact structure, composed of small wiry sticks, coarse grasses, and broad and narrow strips of bark; the inside is lined with the soft woolly portion of the blossoms of small ground plants: it usually lays two, but occasionally three eggs, which are of a pale buff, thinly spotted and freckled with deep chestnut-brown, particularly at the larger end, where they not unfrequently assume the form of a zone; their medium length is nine lines and a half, and breadth nearly seven lines.

Sydney　　　　　**John Gould**[42]

Honeyeaters' food　　　　　**1860**

The New Holland Creeper are seen in great numbers, adorned with gay-coloured plumage, seeking their food, of insects and sweets, amidst the Banksias, Aloes, and festoons of creeping plants. The massive flowering stems of the Aloe, and its numerous blossoms, secrete a honey-like substance; these, with the long spikes of the Grass-tree (*Xanthorrhoea hastilis*), profusely covered with honey-dew, all yield sustenance to this beautiful tribe of birds.

Sydney　　　　　**George Bennett**[11]

The brown colour on the mantle of this specimen is not characteristic of the female; individuals of both sexes may have this marking. Adults are 165–185 mm long. (From White[115])

Great bowerbird
and fawn-breasted bowerbird

Chlamydera nuchalis
C. cerviniventris

The behaviour of bowerbirds has been a source of wonder since their bowers were first discovered. The male builds and decorates a bower for use as a courtship and mating area, which he then decorates with objects such as flowers, berries, feathers, shells and, recently, pieces of plastic and metal.

The extracts reproduced here concern two species: the great bowerbird, so called because of its bright rose-pink nuchal collar and large size; and the fawn-breasted bowerbird, which is found also in New Guinea. The latter is the 'new species' referred to in John MacGillivray's account.

Bowerbirds are closely related to birds of paradise, but whereas the male birds of paradise attract females by means of their bright feathers, bowerbirds use a more architectural approach.

A bird's playhouse   20 November 1843

I found matter for conjecture in noticing a number of twigs with their ends stuck into the ground, which was strewed over with shells, and their tops brought together so as to form a small bower; this was 2½ feet long, 1½ foot wide at either end. It was not until my next visit to Port Essington that I thought this anything but some Australian mother's toy to amuse her child: there I was asked, one day, to go and see the "bird's playhouse", when I immediately recognised the same kind of construction I had seen at the Victoria River: the bird was amusing itself by flying backwards and forwards, taking a shell alternately from each side, and carrying it through the archway in its mouth.

Victoria River, N.T.   **J. L. Stokes**[103]

First sighting of a new species   1849

Two days before we left Cape York I was told that some bower birds had been seen in a thicket, or patch of low scrub, half a mile from the beach, and after a long search I found a recently constructed bower four feet long and eighteen inches high . . . Next morning I was landed before daylight and proceeded to the place . . . taking with us a large board on which to carry off the bower as a specimen . . . While watching in the scrub I caught several glimpses of the *Tewinya*

(the native name) as it darted through the bushes in the neighbourhood of the bower, announcing its presence by an occasional loud churr-r-r, and imitating the notes of various other birds, especially the leatherhead . . . My bower-bird proved to be a new species, since described by Mr. Gould as *Chlamydera cerviniventris*, and the bower is exhibited in the British Museum.

Cape York, Qld   **John MacGillivray**[77]

Gould's description   1869

If any one circumstance more than another would tend to hand down the name of the author of the "Birds of Australia" to posterity, it would be the discovery and publication of the singular habits of the Bower birds . . . The discovery of the present species is due to Mr. John MacGillivray . . . The bower . . . is about 13 inches long and 10 or 11 inches high; its walls, which are very thick, are nearly upright, or but little inclining towards each other at the top, so that the passage through is very narrow. This elevated structure, which is formed of fine twigs, is placed on a very thick platform of thicker twigs, nearly 4 feet in length and almost as much in breadth; here and there a small snail shell or berry is dropped by way of decoration.

Australia   **John Gould**[42]

This woodcut is from John MacGillivray's *Voyage of the H.M.S. Rattlesnake* (published 1852)[77]; it accompanied his description of the first sighting of the fawn-breasted bowerbird.

From Gould's *Birds of Australia*[42], a male great bowerbird in his bower, with his collection of bones and shells. Beneath the bushes in the background is the bower of a rival male.

Regent bird

Sericulus chrysocephalus

The male regent bird is the most brilliantly plumaged of all Australian bowerbirds. Early naturalists thought of it as a honeyeater (see Lewin's account), an oriole (see Lesson's account), and then as a bird of paradise, before it was recognised as a bowerbird.

The male's bower of twigs, built to court the brown-coloured female, is usually well hidden in vegetation on the forest floor. Like all bowerbirds, the male is polygamous and takes no part in nest-building, the incubation of eggs or the raising of young.

It is found from Mackay in central Queensland south to the Sydney district in New South Wales — being common in coastal forests in the Gosford – Wyong area. It eats both native and cultivated fruits, and occasionally insects.

From John Lewin's *Natural History of the Birds of New South Wales* (1822).[73] The bird's plumage is in the black and gold colours traditionally worn in Europe by a monarch's regent.

A perfect regent bird's bower, sighted in 1970 in the Iluka Forest, New South Wales.[37] The usual size is about 260 mm wide x 220 mm long, with walls 200-300 mm high — depending on materials used; the avenue is about 100 mm wide.

'King honeysucker'   1822
LENGTH: Nine inches and a half; bill one inch in length, and of a bright yellow-orange; eye yellow; forehead, crown, and back part of the neck, bright golden-yellow verging to orange; the feathers short appearing like velvet; from the base of the bill to the eye black; above the eye the same; chin, cheek, throat, black; breast, belly, and vent the same; back and shoulders deep shining blue-black; bastard-wing black; primaries the same; secondaries bright golden-yellow, partly tipt with black; tail black and a little forked; legs and claws black. INHABITS: The banks of Patterson's river; frequents thick brushy woods. REMARKS: This beautiful species was shot about 30 miles from the settlement of Newcastle.
Near Newcastle, N.S.W.   **John Lewin**[73]

'L'oriot prince-regent' 1824

The carriage of this bird is indeed entirely that of an oriole, but its tongue, according to what I was told by Mr. Fenton, senior assistant-surgeon to the forty-eight Infantry Regiment, who has dissected several of them, terminates in a brush. This structure appears to have been given to several types of New Holland birds, and their ecology will thus be accommodated to the way of life imposed upon them by necessity, that of sucking the flowers or the nectars of the trees of the forests. This habit appears to be found in a great number of birds of New South Wales, and it is the same among various parrots. This oriole, although not rare at Sydney, is sold there at very high prices, as it is very popular among the English. We brought away a superb individual which is deposited in the Museum.

Sydney **René Lesson**[72]

Males fight 1842

I must mention that at least fifty out of colour may be observed to one fully-plumaged male, which when adorned in its gorgeous livery of golden yellow and deep velvety black exhibits an extreme shyness of disposition, as if conscious that its beauty, rendering it a conspicuous object, might lead to its destruction; it is usually therefore very quiet in its actions, and mostly resorts to the topmost branches of the trees; but when two gay-coloured males encounter each other, frequent conflicts take place.

Maitland, N.S.W. **John Gould**[42]

Further details 1860

The Regent-bird, or King Honeysucker of the colonists has a wide range, from Illawarra (in which district they have occasionally been very numerous), to Port Macquarie, Moreton Bay, and the Clarence River districts. The adult male in full feather is of a golden-yellow colour, beautifully contrasted with a deep velvety black; but the young males resemble the female in their plain, simple plumage.

Sydney **George Bennett**[11]

The difference between male and female colouration of the regent bird is clearly shown here, in Gould's *Birds of Australia*.[42] The birds are feeding on the fruits of a native fig.

Laughing kookaburra

Dacelo novaeguineae

Famous for its raucous laughter-like call, often heard in the early morning, the laughing kookaburra is the world's largest kingfisher (460 mm long, including 65-mm bill). Several of these birds were caught by Joseph Banks in 1770 and on his way home, at the Cape of Good Hope, he gave a specimen to Pierre Sonnerat, who illustrated it in his book *Voyage à la Nouvelle Guinée*, with a caption indicating that the bird came from New Guinea. When Latham described Banks's remaining specimens in 1782 in *A General Synopsis of Birds* [69] he perpetuated Sonnerat's error; this explains how the bird obtained its early scientific name *Dacelo novaeguineae*, even though it does not occur there.

The bird is found throughout woodland and open forest in eastern Australia, with the exception of Cape York, and has been introduced to south-western Australia and Tasmania. Its diet consists mainly of insects but it also eats snakes, lizards and small birds.

From *Voyage à la Nouvelle Guinée* (1776). [101] Sonnerat had accompanied the *Ile de France* and *Le Nécessaire* on explorations of the Philippines and the lands of the Papuans.

This more life-like kookaburra is from Phillip's *Voyage to Botany Bay* (1789).[90]

The bushman's clock   c.1805
The settlers call this bird the *Laughing Jackass* and the natives, as I think, *Cuck'unda*. It is common throughout the colony, at least in all the forest land of the interior parts. It makes a loud noise somewhat like laughing which may be heard at a considerable distance, from which circumstance, and its uncouth appearance, it probably received the above extraordinary appellation from the settlers on their first arrival in the colony. I have also heard it called the Hawkesbury Clock (clocks being at the period of my residence scarce articles in the colony, there not being one perhaps in the whole Hawkesbury settlement) for it is amongst the first of the feathered tribe which announces the approach of day. When sleeping in the woods I have often found its singular voice most welcome in the morning.
Hawkesbury, N.S.W.   George Caley[25]

Distribution of kookaburras   1846
The Great Brown Kingfisher does not inhabit Van Diemen's Land, nor has it yet been met with in Western Australia; it may be said to be almost solely confined to that portion of Australia lying between Spencer's Gulf and Moreton Bay, the south-eastern corner, as it were, of the continent . . . It never visits New Guinea nor even the northern coast of Australia, where its place is supplied by [blue-winged kookaburras]. Unlike most other species, it frequents every variety of situation; the luxuriant brushes stretching along the coast, the more thinly-timbered forest, the belts of trees, studding the parched plains and the brushes of the higher ranges being alike favoured with its presence. Over all these localities it is rather thinly dispersed being nowhere very numerous.
Eastern Australia   John Gould[42]

Kookaburras preying on snakes   1834
A gentleman told me that he was perfectly aware of the bird destroying snakes, as he had often seen them carry the reptiles to a tree, and break their heads to pieces with their sharp strong beaks . . . It is not uncommon to see these birds fly up with a long snake pending from their beak, the bird holding the reptile by the neck, just behind the head; but as the snake hangs down without motion, and appears dead, it is probable that the bird destroys them upon the ground before it conveys them into the tree.
New South Wales   George Bennett[10]

Reptiles

The class Reptilia is divided into four orders, three of which occur in Australia: crocodiles, turtles, and lizards and snakes. The fourth order is represented only by the tuatara of New Zealand. In this book only the lizards and snakes are included.

Lizards and snakes reproduce either by laying eggs (oviparity) or by giving birth to live young (viviparity), or a combination of both where the egg is retained and hatches inside the mother (a condition called ovoviviparity).

Goannas are members of a family elsewhere called monitor lizards; this family includes the world's largest lizard, the 3-metre-long komodo dragon of Indonesia, and the slightly smaller perentie of the Australian deserts. Of the goannas in this book, some individuals may attain a length of 2 metres.

The picturesque frilled lizard appears on the Australian two-cent coin. Its colour varies from grey to brown or orange-brown, with yellow to black frill, often variegated with orange. A related species, but looking entirely different, is the thorny devil; of bizarre appearance, this lizard lives in the great central and western deserts. Both the frilled lizard and the thorny devil lay eggs.

The family of skinks is the largest family of Australian lizards, and the blue-tongued lizard and stumpy-tailed lizard are two examples. Both are viviparous.

The greatest number of species of snakes in Australia are front-fanged venomous snakes (most snakes elsewhere are back-fanged). The majority of Australian snakes are only mildly venomous – though the brown snake is a species potentially dangerous. It is found over all of the Australian mainland. The beautifully coloured red-bellied black snake is a common reptile around swamps, lagoons and streams.

Diamond pythons are members of a group that includes the world's largest snakes. Pythons kill their prey by squeezing until suffocation occurs – and the diamond python is commonly kept by farmers in hay sheds to catch rats and mice.

Goannas

Varanus spp.

Goannas are found all over Australia except Tasmania. There are more than fifteen different species, the largest of which may grow to 2.5 metres long. Some are tree-dwelling.

These lizards eat meat, including rats, mice, rabbits, insects and snakes, and particularly nesting birds. Small specimens are rarely seen because they are well camouflaged and probably naturally cautious until they grow sufficiently strong to win a fight with their enemies – hawks, kookaburras and people.

The most common goanna, *Varanus gouldii*,

This goanna with its forked protruding tongue is from Phillip's *Voyage to Botany Bay* (1789).[90]

White's illustration of the 'Variegated Lizard' stalking through swamp vegetation shows a tail much longer than would be found naturally.[115]

rarely climbs trees, but some species climb well and when cornered may even run up a human being by mistake.

The European explorers called them guanas, from the Spanish *inguana*; this term (which correctly refers to a different group of lizards) has become corrupted to 'goanna'.

First description 1789

This most elegant species is in length, from the nose to the end of the tail, about forty inches . . . the tongue is long and forked . . . This beautiful lizard is not uncommon at Port Jackson, where it is reputed to be a harmless species. Individuals vary much one from another, in respect to the length of the tail, as also in the colour of the markings; some having those parts marked with a pure silvery white . . .

Port Jackson **Arthur Phillip**[90]

Variegated colour 1790

This Lizard approaches so extremely near to the Lacerta Monitor of Linnaeus, or Monitory Lizard, as to make it doubtful whether it be not in reality a variety of that species. The body is about 15 inches in length, and the tail is considerably longer. The animal is of a black colour, variegated with yellow marks and streaks of different shapes, and running in a transverse direction. On the legs are rows of transverse round spots; and on the tail broad alternate bars of black and yellow. In some specimens the yellow was much paler than in others, and nearly whitish.

Port Jackson **John White**[115]

Goannas as food 10 September 1836

Natives are extremely fond of the Guana or Iguana, a reptile of the lizard kind, but they do not attain so great a size in New South Wales as in Africa or America; I have killed them, however, as long as five feet. When surprised they will immediately take to a tree, but a black will always ascend after them. On one occasion, a native, who accompanied me, killed a she goana, and carried it as far as five miles to the tents, when he immediately commenced eating it. He took from it a string of eggs, about thirteen in number, which were as large as pigeons eggs; these he put under the ashes, and soon after commenced eating them; I also ate two or three, and they were delicious. The bite of the Iguana is severe but not venomous, and they have been known to fasten to a lamb or sheep, and suck their blood.

New South Wales **W. R. Govett**[45]

A finely marked goanna – perhaps one that has just shed its skin – reproduced in the *Proceedings of the Zoological Society of London* and cut for the scrapbook compiled by Albert Gunther, Keeper of Fishers at the British Museum (Natural History) during the late nineteenth century.[54]

Thorny devil

Moloch horridus

This excellent engraving was published with the first description by J. E. Gray, a keeper of the British Museum, in Captain George Grey's *Journals of Two Expeditions of Discovery in North-west and Western Australia during the Years 1837, 1838 and 1839*.[48]

The thorny devil lives in the great central and western deserts. It is able to absorb water from the ground by tiny capillary channels on its skin which lead to the mouth; when the water reaches the mouth it is swallowed in the normal manner.

It eats tiny black ants; when feeding it consumes one ant or more every two seconds. It is therefore difficult to feed these animals in captivity, and most die. (They are a protected animal under most National Parks and Wildlife Acts.)

When attacked, it protects its head between its legs and the hump on its neck is pulled forward to look like a head.

First description — 1841

The external appearance of this Lizard is the most ferocious of any that I know, the horns of the head and the numerous spines on the body giving it a most formidable aspect . . . I have named this genus, from its appearance, after "Moloch, horrid king" . . . This is the Spinous Lizard exhibited by Mr. Gould at the meeting of the Zoological Society [London] in October 1840.

North-western Australia   **J. E. Gray**[47]

Observations of captive specimens — 1867

The three Lizards last received I kept always together in an airy wooden box, with plenty of sand at the bottom . . . All attempts to make them eat were ineffectual, as before remarked. They were supplied with living and dead insects (which are said to form their food) both day and night. Sugar &c. also was given, in case they should, as so many animals do, fancy sweet things; but each and all appeared to be regarded with the same indifference as everything else . . . Among the facts ascertained from observations and reliable information, one is that these Lizards are perfectly harmless; that is, on no occasion do they attempt to bite or scratch; and with this knowledge I have continually taken them up by hand and examined them. The mouth of the animal is very small; and it is apparently unable to bite; and I have no reason to think that, if it did so, any unpleasant consequences would follow.

The animals changed colour frequently while I had them, from their original bright hues to a dull slate- or soot-colour, under which their markings were but dimly seen . . . The change of hue never occurred suddenly, or while the creature was being looked at; but after a long interval (say, after a day or a night had passed) I observed the alteration . . . They are diurnal reptiles only — that is, not abroad at night . . . The eye is not suited for nocturnal excursions, being small and deeply set.

South Australia   **C. A. Wilson**[118]

A scientific description — 1885

Mouth small; lateral teeth in the upper jaw implanted horizontally, directed inwards. Tympanum distinct. Body depressed. Tail short, rounded. Upper and lower parts covered with small scales or tubercles intermixed with large spinous tubercles. Nape with a large roundish protuberance. No femoral or praeanal pores.

Central and western Australia   **Albert Gunther**[53]

Frilled lizard

Chlamydosaurus kingi

The frill of this lizard is raised by soft bone-like structures that extend from the tongue into the frill; when the mouth is opened wide, the frill raises automatically. Used to deter enemies, the frill display also functions in courtship.

First sighting   8 October 1820

I secured a lizard of extraordinary appearance, which had perched itself upon the stem of a small decayed tree. It had a curious crenated membrane like a ruff or tippet around its neck, which it spreads five inches in the form of an open umbrella.

Careening Bay,
Port Nelson, W.A.   **Allan Cunningham**[63]

From Albert Gunther's scrapbook, Illustrations Relating to Australian Fauna: 'Australian Reptiles 1789-1897'.[54] The frilled lizard is distributed across northern Australia.

Another early sighting   1841

As we were pursuing our route in the afternoon, we fell in with a specimen of the remarkable frilled lizard (*Chlamydosaurus Kingii*); this animal measures about twenty-four inches from the tip of the nose to the point of the tail, and lives principally in trees, although it can run very swiftly along the ground: when not provoked or disturbed it moves quietly about, with its frill lying back in plaits upon the body: but it is very irascible, and directly it is frightened, elevates the frill or ruff, and makes for a tree; where, if overtaken, it throws itself upon its stern, raising its head and chest as high as it can upon the forelegs, then doubling its tail underneath the body, and displaying a very formidable set of teeth, from the concavity of its large frill, it boldly faces any opponent, biting fiercely whatever is presented to it, and even venturing so far in its rage as to fairly make a fierce charge at its enemy. We repeatedly tried the courage of this lizard, and it

certainly fought bravely whenever attacked. From the animal making so much use of this frill as a covering and as means of defence for its body, this is probably one of the uses to which nature intended the appendage should be applied.

North-western Australia **George Grey**[48]

An illustration by Prêtre of a frilled lizard in typical threat display, from *Erpetologie Génèrale* (1854) by A. Dumeril, adminstrator of the Paris Natural History Museum.[35]

Another description 1883

On one of the two occasions on which I have seen the lizard adopt its biped mode of locomotion, trotting out briskly on its hind legs, its fore-paws hanging down affectedly and its vertebral line to the very snout stiffened at an angle of 60°, I was much interested to see it halt abruptly, erect its frill, and at the same moment turn its head enquiringly from side to side – then trot on again for twenty yards or so, and repeat its attitude of attention – thus it did, till it reached the tree it was making for, then darting a few feet up its bole it clung there immovable for more hours than my leisure could afford for observation.

Queensland **Charles Walter de Vis**[34]

Blue-tongued lizard

Tiliqua scincoides

This lizard turns to face an attacker, opens its mouth wide, and pokes out its blue tongue. Simultaneously, it inflates its body with air, and hisses. It is not poisonous, although it has a strong bite.

One of Australia's largest skinks, at 450-500 mm, the blue-tongued lizard ranges from the south-east of the continent, along the east coast and across northern Australia. The young are born alive and as many as twenty-five may be produced at once – thus explaining its commonness. It is a useful garden pet, for it eats snails and will clear up scraps of meat and fruit. If not disturbed, it will live for many months in the same home, under a pile of stones or in a drainpipe.

The 'scincoid' lizard drawn by F. P. Nodder for Shaw's *Naturalist's Miscellany*[97] is a mirror image of the first illustration in White's journal (see illustration on p. 98). Nodder was the artist for both works.

The striped body is well camouflaged in the Australian bush – the colours vary between individuals and habitats.

First description   **1790**

This Lizard comes nearer to the Scincus than any I am acquainted with, but is still a distinct species . . . The tail is longer than that of the Scincuses, and not so taper; the animal is of a dark iron-grey colour, which is of different shades in different parts, forming a kind of stripe across the back and tail. The scales of the cuticle are strong, covered with the same kind of scales as the body, but the scales of the feet are not. On the cuticle are small knobs as if it were studded.

The toes on each foot are pretty regular; the difference in length not great, and the same on both the fore and hind foot; which is not the case with the Scincus, it having a long middle toe. There are small short nails on each toe; on their upper surface they are covered with a series of scales, which go half round, like a coat of mail.

The teeth are in a row on each side of each jaw, becoming gradually larger backwards. They are short above the gum, and rounded off, fitted for breaking or bruising of substances, more than cutting or tearing.

New South Wales   **John White**[115]

From the scrapbook compiled by Albert Gunther of Illustrations Relating to Australian Fauna: 'Australian Reptiles 1789–1897'.[54]

Another engraving from Gunther's scrapbook. These illustrations originally appeared in the *Proceedings of the Zoological Society of London*.[54]

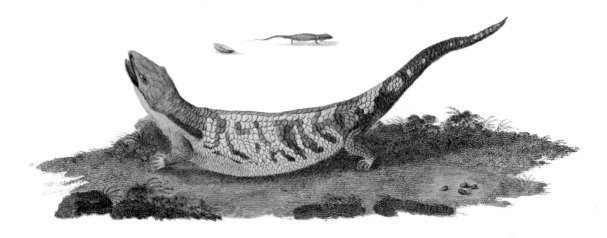

An early description　　　　**June 1794**

The tongue is not forked, as in the generality of lizards, but is broad, flat, and rounded. The teeth rather small than large, somewhat obtuse, and short. The colour of the animal is a pale yellowish brown, with a long patch or mark of very deep brown or blackish on each side of the neck. The sides are also tinged with the same colour, and the tail more deeply so than the sides. In a much smaller specimen of the same animal the tail had the appearance of being fasciated with a number of transverse bars of deep brown. In some specimens the tail, towards the end, is furnished with a sort of process or appendage, resembling a secondary tail, as it were, and it has been imagined that this might be a sexual distinction; and perhaps peculiar to the male: but it seems infinitely more probable that such an appearance is owing to mere accident; it being a well-known fact that the tails

A blue-tongued lizard in the journal of John White, chief surgeon with the First Fleet, illustrated with . . . *Sixty-five Plates of Non descript Animals, Birds, Lizards, Serpents, curious cones of Trees and other Natural Productions.*[115]

of lizards, when injured, or mutilated by accidental violence, frequently reproduce in different directions; of which numerous examples may be seen in the works of natural historians. The Scincoid Lizard is a native of New Holland, and is commonly about eighteen inches in length.

New Holland　　　　**George Shaw**[97]

Stumpy-tailed lizard

Trachydosaurus rugosus

First sighting **6 August 1699**

And a Sort of Guano's, of the same Shape and Size with other Guano's, describ'd but differing from them in 3 remarkable Particulars: For these had a larger and uglier Head, and had no Tail: And at the Rump, instead of the Tail there, they had a Stump of a Tail, which appear'd like another Head; but not really such, being without Mouth or Eyes: Yet this Creature seem'd by this Means to have a Head at each End; and, which may be reckon'd a fourth Difference, the Legs also seem'd all 4 of them to be Fore-legs, being all alike in Shape and Length, and seeming by the Joints and Bending to be made as if they were to go indifferently either Head or Tail foremost. They were speckled black and yellow like Toads, and had Scales or Knobs on their Backs like those of Crocodiles, plated on to the Skin, or stuck into it, as part of the Skin. They are very slow in Motion; and when a Man comes nigh them they will stand still and hiss, not endeavouring to get away. Their Livers are also spotted black and yellow: And the Body when opened hath a very unsavoury Smell. I did never see such ugly Creatures any where but here. The Guano's I have observ'd to be very good Meat: And I have often eaten of them with Pleasure; but tho' I have eaten of Snakes, Crocodiles and Allegators, and many Creatures that look frightfully enough, and there are but few I should have been afraid to eat of, if prest by Hunger, yet I think my Stomach would scarce have serv'd to venture upon these N. Holland Guano's, both the Looks and the Smell of them being so offensive.

Shark Bay, W.A. **William Dampier[30]**

Many of the details noted by Dampier 160 years earlier are illustrated in this plate from a Victorian scientific journal, which Gunther included in his scrapbook[54] (now in the Mitchell Library, Sydney).

First scientific description 1827

The body nearly uniform, chestnut brown; the head depressed with the scales convex, and more nearly of an equal size than usual: those round the eyes and mouth large; the three anterior scales on the edge of the lower jaw larger than those which cover the lower surface of the head, body, and tail, which are uniform, distinct, large and membranaceous: the scales of the back are nearly of equal size with those covering the commencement of the tail; they are furnished with a prominent midrib, and end in a point. The legs very short, compressed covered with nearly smooth, rather thin scales. The toes very short; claws rather thick, and short. The tail about half the length of the body. Head three inches long. Body, seven inches. Tail, four inches. Only one specimen of this exceedingly interesting animal was brought home by Captain King, but the spirits in which it had been preserved had unfortunately evaporated, so that it was considerably injured . . . The above specimen was found at King George the Third's Sound, and is preserved in the Museum [the British Museum].

George's Sound, W.A. J. E. Gray[46]

Dampier's fascinating description of the stumpy-tailed lizard, which he called a guano, is surprisingly accurate. The tail is shaped similarly to the head; it is full of fat and acts as a reserve food supply. In winter or during a dry summer, the lizard can fast for weeks.

Each female produces one, two or rarely three young. The young are born alive, hatching from eggs inside the mother. (Most reptiles hatch after the mother has deposited the eggs.) The diet includes insects, snails, wildflowers, fruit and berries.

The colour of the stumpy-tailed lizard varies in different districts; it is found all over the southern half of the mainland but not in Tasmania, and grows to 300 mm.

Diamond python

Morelia spilotes

The diamond python is a member of the family that includes the world's largest snakes – the amethystine or rock python has been recorded as exceeding 8 metres in length – but the diamond grows usually only to 2 metres, with 4 metres a maximum. It is found only in the coastal forests of New South Wales.

Pythons kill their prey by squeezing until it suffocates; the prey is then swallowed whole, the jaws becoming disconnected at the corner of the mouth and the skin extending. Ten to twenty eggs are laid, and, uniquely among the snakes, the pythons coil around the eggs to incubate them.

The plate that accompanied John White's description.[115] It seems to have been drawn from a preserved specimen.

First description   1790

Snake, No. 5, upwards of eight feet in length, of a blackish colour, varied with spots and marks of a dull yellow: the belly also is of a yellowish colour. The scales are small in proportion to the size of the animal; the tail gradually tapers to a point.

Sydney   **John White[115]**

A python from White's journal (published 1790).[115] The illustration is scientifically inaccurate – a snake does not move with its body thrown in vertical loops – but it does capture the impression of the python's speed.

The Aboriginal name   1860

The Diamond-Snake (*Morelia spilotis*), one of the Boa tribe, is not venomous, and feeds upon opossums, rats, mice, and birds; it attains a length of from 12 to 15 feet. It is very handsome, and is called *Kurrewa* by the natives of Port Macquarie. The scales of the back are diamond-shaped, and the whole back dotted with bright yellow upon a deep purplish ground; the abdomen is of a light yellow or straw colour. The Diamond-Snakes soon become very tame: I permitted one, about 8 feet long, to entwine around my arm; but I found its pressure (which it seemed to exercise merely by the muscular power necessary to retain its position) was sufficiently energetic to make my arm ache for some hours afterwards.

New South Wales   **George Bennett[11]**

How it kills
1870

A medical friend up country once heard a great noise in his stable, and on opening the corn bin, he found a large diamond snake coiled four times around a large cat, and killing it by constriction only, leaving no doubt about the *modus operandi* of killing its prey.
New South Wales   **Dr Berncastle**[12]

A lifelike lithograph of the diamond python coiled in a blueberry ash tree, perhaps waiting for a bird or possum, from Gerard Krefft's *Snakes of Australia* (published in Sydney in 1869).[65] Krefft was secretary, then curator, of the Australian Museum in Sydney and was a recognised authority on snakes. His duties went far beyond the purely scientific, however, and an early handbill advertising the museum – which was open between 7 p.m. and 10 p.m. – mentioned that 'Mr. Krefft will be in attendance and give every possible information to visitors'. They had on display 'Native Bears . . . very tame and exceedingly funny'.

Red-bellied black snake

Pseudechis porphyriacus

The red-bellied black snake is a common reptile around swamps, lagoons and streams throughout south-eastern Australia. It feeds mainly on frogs and small mammals and grows to a length of 2 metres. Although classified as venomous, its bite is usually not serious, but it is painful. The venom destroys the walls of small blood vessels in the locality of the bite, causing internal bleeding and swelling. The snake will retire rather than attack, unless it is provoked.

The young are born in sacs which rupture within a few minutes after birth; between eight and forty are born.

First described
1790

This beautiful Snake, which appears to be unprovided with tubular teeth or fangs, and consequently not of a poisonous nature, is about three, and sometimes four feet in length. Its colour on the upper parts is a glossy violet-black; the sides of an elegant deep crimson, which on the abdomen declines into a paler tinge, or more approaching to whitish; while the scuta or broad semi-circular scales which compose this part, are each deeply bordered with black. The alternate scales or divided scuta beneath the tail are of a lead-colour, and the largest series of the crimson side-scales are tipped with black; which gives them an appearance peculiarly elegant. This Snake may be considered as a species hitherto undescribed.
New Holland   **George Shaw**[98]

The fangs observed
1801

The most formidable among the reptiles was the black snake with venomous fangs, and so much in colour resembling a burnt stick, that a close inspection only could detect the difference. Mr. Bass [George Bass, who circumnavigated Tasmania with Flinders] once, with his eyes cautiously directed towards the ground stepped over one which was lying asleep among some black sticks, and would have passed on without observing it, had not its rustling and loud hiss attracted his attention the moment afterwards.
Derwent River,
Van Diemen's Land   **David Collins**[6]

A black snake from *Zoology and Botany of New Holland* (1793) by George Shaw, a keeper of the British Museum.[98] For the journey to England, specimens were often preserved in alcohol, which tended to extract the colours from them.

More notes   1860

Four-fifths of the Snakes as yet sent from different parts of Australia are poisonous, and many very virulent. The Black Snake (*Pseudechis porphyriacus*) is common over the colony, and is principally met with in marshy places or near to water. It is of a beautiful glossy black over the back, and blood-red over the abdomen. It measures from 5 to 8 feet in length. The poison-fangs are small.

Sydney   **George Bennett**[11]

The snake's venom   1892

But although the yield of poison is small, the virulence of our Black snake compares very favourably with that of the Cobra. Some idea of this may be gathered from the fact that 1/1000th. grain invariably kills a rabbit of 5 lbs weight, in about one hundred seconds.

Sydney   **C. J. Martin**[80]

Snakes are often drawn coiled to show the under surface, sides and top simultaneously; the pattern of scales and the colour are important characteristics for identification. The undersurface of a red-bellied black snake varies in colour from red to white, except at the tip of the tail, which is invariably black. From Krefft's *Snakes of Australia*.[65]

Brown snake

Pseudonaja textilis

The brown snake has a potent venom and, although it does not attack unless provoked, when roused it may strike repeatedly. At each bite, a small quantity of highly poisonous venom is automatically ejected. The venom is stored in a pair of glands behind the eyes; when the mouth closes on the flesh of the victim these glands are squeezed and the poison ejected through a pair of hollow teeth at the front of the upper jaw. The venom paralyses the muscles of the lungs and heart of the prey. Before a brown snake strikes, it coils its neck into an s-shape.

Found over all of Australia except Tasmania, it feeds on small mammals and reptiles; and by its range inhabits a wide variety of environments. The female lays ten to thirty-five eggs.

An early description   1834

The "brown snake", which I examined, is also venomous, and, according to popular opinion, the effect very dangerous upon the human constitution. The specimen measured nearly five feet in length, and five inches at its largest circumference; the upper part of the body was of a brown colour (from which no doubt its name is derived), with a few light shades of black; the abdomen was of a light bluish black. In the stomach were found several half-digested lizards, and a quantity of worms, which in some parts had even perforated the coats: on a further examination, the lungs were also found perforated by, and had attached to them, a number of these worms, varying from one and a half to two inches in length, and of a bright red colour: I preserved them, together with the lungs, in spirits and sent them to the museum of the Royal College of Surgeons, in London.

New South Wales   **George Bennett[10]**

More details   29 October 1836

The Grey and the Brown Snakes, which are also numerous, are found near the river banks, and sometimes on open plains. They, too, are said by the natives to be dearly in their bite.

Argyle County, N.S.W.   **W. R. Govett[45]**

Snakes are difficult to illustrate because of their length; usually 1.5 metres long, the brown snake from Krefft's *Snakes of Australia*[65] is coiled to save space. The picture is technically correct, and Krefft records in his preface that 'the gifted daughters of A. W. Scott Esq. M.A. have done everything in their power to give correct figures of the reptiles illustrated'. Harriet Scott drew the snake from a photograph of a specimen skewered to keep it in position.

Fish

Fish are by far the most numerous species of vertebrates, outnumbering the combined totals of birds, mammals, reptiles and amphibians. They are divided into three classes: the jawless fish, the cartilaginous fish (sharks, rays and skates), and the bony fish. The latter class is further subdivided into fleshy finned fish (the coelacanth and the lungfish) and the ray-finned fish, so named because breast and pelvic fins are stiffened by bony rays. Some 2 000 species in three of these four groups occur in Australian waters.

The class that includes sharks and rays is called cartilaginous because their skeletons are not bony, but flexible cartilage – a material found only in the young of most species and in small areas, such as joints, of other adult vertebrates. Some of the Australian sharks are zoologically interesting, for only fossil teeth of similar species have been found in other parts of the world. However, all sharks are 'living fossils'; they evolved in Devonian times (350 million years ago) and have survived virtually unchanged since.

The Australian lungfish, too, is a 'living fossil'. It has a modified air-bladder to act as a lung, enabling it to survive in highly deoxygenated water. The coelacanth was thought to be a fossil until discovered just before World War II in African waters. Few have since been found – none near Australia – although a specimen is on display in The Australian Museum in Sydney.

In contrast to the vast array of marine species, the Australian freshwater species are few: there are only some 200 species, and of these a large proportion were derived from marine fish that became adapted to estuarine conditions and thence to freshwater.

Some freshwater species are very important economically and as game animals. The Murray cod is confined to the Murray-Darling river system, where it may grow to enormous size. The golden perch is found in these rivers but also in the Dawson and Mary Rivers in Queensland – thus separated by the Great Dividing Range. This discontinuity in distribution is puzzling.

Port Jackson shark

Heterodontus portusjacksoni

The Port Jackson shark grows to a length of only 1.5 metres. It has grinding teeth – its diet of sea urchins and starfish stains the mouth and teeth deep violet or red – and it does not attack people for food.

The egg case is dark brown and pear-shaped, surrounded with spiral flanges. It is usually deposited in a crevice in rocks or jammed between marine growths, and contains one embryo. (Many people may have seen the egg case washed up on a Sydney beach – after the young has hatched – but mistaken it for seaweed.) The young take several months to develop in the egg, and when they hatch are able to fend for themselves immediately.

The shark is found in moderately shallow coastal waters, ranging from New South Wales waters around the south-eastern mainland to the south-west.

From Arthur Phillip's *Voyage to Botany Bay*, published in London in 1789, one year after he arrived as first governor.[90] The detail of the jaw shows the development and growth of the rows of teeth.

First description   **1789**

The length of the specimen from which the drawing was taken, is two feet; and it is about five inches and an half over at the broadest part, from thence tapering to the tail: the skin is rough, and the colour, in general, brown, palest on the under parts: over the eyes on each side is a prominence, or long ridge, of about three inches; under the middle of which the eyes are placed: the teeth are very numerous, there being at least ten or eleven rows; the forward teeth are small and sharp, but as they are placed more backward, they become more blunt and larger, and several rows are quite flat at top, forming a kind of bony palate, somewhat like that of the Wolf-fish; differing, however, in shape, being more inclined to square than round, which they are in that fish: the under jaw is furnished much in the same manner as the upper: the breathing holes are five in number, as is usual in the genus: on the back are two fins, and before each stands a strong spine, much as in the Prickly Hound, or Dog, fish.

Port Jackson   **Arthur Phillip**[90]

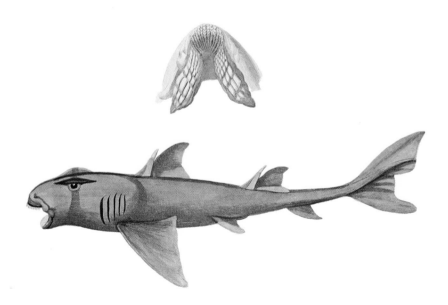

In Port Jackson, Sydney, New South Wales, within the Heads of that harbour, and at present only found in that limited locality, the singular species of Shark, known by the name of the Port Jackson Shark . . . is met with. It does not grow to a large size, seldom attaining a greater length than from 3 to 4 feet. The jaws are armed with strong bony plates, serviceable for the purpose to which they are applied, that is, for grinding down the shell-fish which forms its food. The colour . . . is a sandy-brown, paler underneath: the skin is rough, and very much resembles, with its spines, the "Dog-fish" of the British coasts; indeed, it bears that appellation generally among the Sydney fishermen.

Sydney   **George Bennett**[11]

Wobbegong shark

Orectobolus maculatus

This unusual shark was described from Port Jackson in the year of British settlement (1788) by the French naturalist Bonnaterre. In eighteenth-century writings it was described as a vicious monster, but apart from rare accidents it does not deliberately attack people. It feeds on reef fish, small octopuses and young crayfish; this upsets crayfishermen, as 'wobbies' often force their way into lobster pots and are there stuck because of the size of their heads.

In Bonnaterre's description, he gave the locality as 'La mere du Sud'; more precisely, it is found in rocky-shore waters off every Australian state. An alternative name is the carpet shark, so called because of the mottled skin colouration which enable it to blend into its surroundings. The fleshy appendages of the face, which also assist in camouflage, act as sensors.

The face of a wobbegong, and a full-length engraving of the shark diving in Botany Bay.[90] A carpet shark has a pair of apertures, called spiracles, behind the eyes and connecting to the throat, which deliver a steady stream of oxygenated water to the gills; the shark therefore does not need to swim continually, as do other sharks, but is able to lie disguised on the sea floor. The wobbegong's camouflage is well shown in the drawings of nearly 200 years ago.

First encounter   **2 November 1789**

This fish was met with in *Sydney Cove, Port Jackson*, by *Lieutenant Watts*, and is supposed to be full as voracious as any of the genus, in proportion to its size; for after having lain on the deck for two hours, seemingly quiet, on Mr. Watt's dog passing by, the shark sprung upon it with all the ferocity imaginable, and seized it by the leg; nor could the dog have disengaged himself had not the people near at hand come to his assistance.

Port Jackson, N.S.W.   **Arthur Phillip**[90]

Small for a shark?   **1789**

This species is considered as one of the smallest of its genus, the specimens hitherto observed, having rarely exceeded the length of three feet. It is a native of the Antarctic seas.

Botany Bay   **George Shaw**[97]

A more accurate estimate of size   **1870**

About seven skinny, simple or partly bifid lobes on each side of the head, five of which are near the angle of the mouth. Very minute barbels across the chin are sometimes absent. Distance between the two dorsal fins equal to the length of the base of the first. Upper parts brown, marbled with grey; a whitish spot behind the spiracle. Stuffed, 7 and 5 ½ feet long. South Australia.

Australian seas   **Albert Gunther**[50]

Lungfish

Neoceratodus forsteri

First sightings   **1850s**

It is strange that a curious creature like this, which was well known to the early settlers at Wide Bay and other Queensland districts, should so long have escaped the eyes of those interested in natural history. I remember that Mr. William Forster mentioned a "fish" with cartilaginous backbone years ago, and that I expressed an opinion that he must be mistaken. This animal is excellent eating, has salmon-coloured flesh, and at certain seasons will rise to a fly; so that the northern squatters have named it the Burnett or Dawson Salmon, from its habits and from the rivers in which it is principally found. The poor bush-cooks who dressed these "Salmons" could have made a small fortune, had they preserved the heads and sent them to Sydney.

Burnett and Dawson Rivers, Qld

       A. Gunther[51]

First popular description   **17 January 1870**

One of the most important discoveries in Natural History has lately been made by the Minister for Lands, the Hon. William Forster, M.L.A., in the shape of an amphibious creature inhabiting northern streams and lagoons, the teeth or dental plates of which resemble some fossil fish-teeth of the Triassic Period . . . I have named this strange animal *Ceratodus Forsteri*. A full description, with figures, will be found in the Proceedings of the Zoological Society of London for the current year, and I will now add a short general description for the readers of the *Herald*. Fancy a triton or newt three feet in length, covered with huge scales the size of a crown piece, with four scaly flappers or fins instead of legs, with four plates of horny teeth, two in each jaw, which resemble segments of a cog-wheel in a horizontal position – with a tail like a Siluroid or Catfish, with small eyes and a partly cartilaginous skeleton, and the modern *Ceratodus* is complete. There are two flat incisor teeth in the upper jaw resembling the third incisor of a two-year-old kangaroo – the lower jaw is destitute of incisors.

Some further interesting facts will probably be added when the spirit-preserved viscera will come to hand. I shall be glad to see my geological friends who are interest in fossil fishes, and I think that I can prove to their satisfaction that the position which I have assigned to the genus *Ceratodus* will be found correct.

Burnett and Mary Rivers, Qld   **Gerard Krefft**[66]

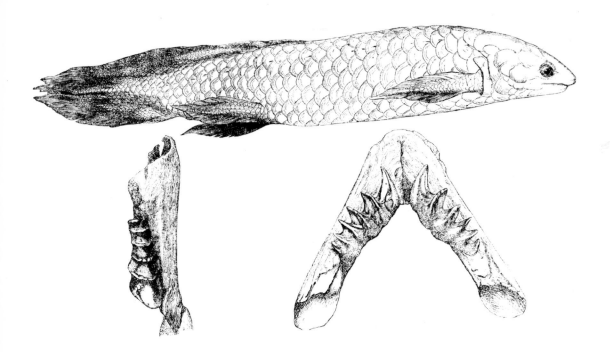

Krefft's illustration of the fish accompanied his paper in the *Proceedings of the Zoological Society of London*.[67] He thought the lungfish was an amphibian, for its spawn resembles that of a frog. The newly hatched young have external gills and resemble newt larvae.

The lungfish is the only living Australian representative of a group common 300 million years ago; only four other distantly related species survive, in Africa and South America.

Its most remarkable feature is the single lung modified from an air-bladder. The fish inhabits the Mary and Burnett river systems in Queensland; and when its river dries to a stagnant waterhole, it can use the lung, instead of the gills, to obtain oxygen.

The Australian species differs from the lungfish found in South Africa and South America in having very large scales and paddle-shaped pelvic and pectoral fins, in addition to an unpaired lung. Unlike them, too, it does not build mud burrows.

The species feeds on small fish, frogs, crustaceans and water snails. The female lays eggs on the bottom mud, and these are guarded by the male. The fish is now totally protected.

Spawning fish observed   **10 October 1964**

I observed several large lungfish swimming about close inshore, over a large weed-covered bank in from two to three feet of water. After watching them for a short while it became obvious that two of the fish were staying close together, the movement of one clearly influencing that of the other. From an elevated rocky ledge six feet above the water we were easily able to view the fishes' activities with the aid of head torches, the light from which did not seem to disturb the fish. Gradually their movements became more restricted in range to a patch of weed about ten feet from the edge of the water. Swimming to and fro, over and through the weed, they gave the appearance of playing "follow the leader". During this time the second fish repeatedly nosed the cloacal region of the leader and was seen to "bump" it several times with its snout. This same fish was seen several times to take in its mouth a long strand of what appeared to be weed, and wave it about. Both fish were then observed from time to time to dive repeatedly through a localized area of weed, often disappearing from view for a few feet. During these "dives", one fish would follow the other closely and both were seen to shake their tails rapidly from side to side . . . the weed where the fish had spent most of their time was searched the following morning, and about 80 eggs were found within a short time.

Barambah Creek, Qld   **Gordon C. Grigg**[49]

Murray cod

Maccullochella peeli

The Murray cod was encountered during Thomas Mitchell's expedition into inland south-eastern Australia; this illustration was drawn for the expedition report, published in London in 1839.[83]

The fish named   24 January 1839
We soon found that this river contained fish in great abundance . . . the fish we had found in the Peel [New South Wales], commonly called by the colonists "the cod", although most erroneously, since it had nothing whatever to do with maleopterygious fishes . . . [It is] light yellow, covered with small irregular display spots, which get more confluent towards the back. Throat pinkish, and belly silvery white. Scales small, and concealed in thick epidermis. Fins obscure. The dorsals confluent. The first dorsal has 11 spines, and the caudal fin is convex.
Murray River   **Thomas Mitchell**[83]

Aboriginal names   1883
Oligorus macquariensis, Cuv. and Val., which is the "Kookoobul" of the Murrumbidgee natives, "Pundy" on the Lower Murray. In this species the height of the body is four times and three quarters in the total length, the length of the head three and a half, the diameter of the eye is one seventh of the latter . . . pectoral and ventral fins short; fifth dorsal spine the longest; second the third spine of the anal fin nearly equal; colour greenish brown, with numerous small dark green spots; belly whitish, but the colour varies much in different places.
New South Wales   **J. E. Tenison-Woods**[119]

A record catch   1 December 1943
It was at Kow Swamp that I landed my specimens of the greatest of all fresh-water fish, known all over Australia as Murray Cod . . . In the morning when I went to get my boat I tried the line and felt something very heavy, like a log, on the end. Thinking the line was snagged, I gave it a sharp pull, and to my surprise there was a large cod on the end. I was afraid the line would break, so had to use all my wits to land the fish . . . After about fifteen minutes struggling I pulled it into the boat . . . the fish weighed 106 pounds. It was four feet nine inches long and thirty-nine inches around the girth.
Kow Swamp, Vic.   **F. J. Marett**[80]

The explorer George William Evans was the first European to see the Murray cod. This fish is well known to Australian fishermen, both Aboriginal and white; its flesh is highly esteemed, having a delicate flavour and tender consistency.

The Murray cod, which is a perch not a cod, inhabits the Murray River and its tributaries. It is also found in the Richmond and Clarence Rivers in New South Wales and in the Dawson and Mary Rivers in Queensland; it is now artificially propagated by Fisheries Departments in those states and in Victoria.

The male is said to brush away surface mud using its fins to create a depression into which a female lays eggs; the male then deposits milt on to the eggs and guards them until they hatch. The fish is more active in summer months and at night.

Diet consists of freshwater crustaceans, mussels, small fish and frogs. It is said to find centipedes irresistible.

Golden perch

Plectroplites ambiguus

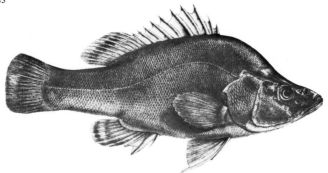

Albert Gunther, Keeper of Fishes in the British Museum of Natural History during the late nineteenth century, wrote the description in scientific terminology for the *Proceedings of the Zoological Society of London*. It is included here as an example of some of the technical first descriptions that were rejected when compiling this book; others are in Latin, which at one time was the universal scientific language.

The golden perch (also known in New South Wales as the yellowbelly) inhabits the Murray-Darling River system, and the Dawson and Mary Rivers of Queensland. It is sometimes called a 'stillwater' fish because of its abundance in lagoons and billabongs. The adult weighs some 4 kilograms — although one specimen taken in 1938 from the Kow Swamp in Victoria weighed 24.5 kilograms gutted. The fish is carnivorous, eating freshwater crustaceans and mussels.

Description of a 'new' fish   12 April 1871
The British Museum has recently received an example of an undescribed Percoid Fish from the Macquarie River, which although allied to *Lates* and *Oligorus* may be regarded as the type of a distinct genus, to be characterised thus:
Ctenolates Seven branchiostegals; pseudobranchiae well developed. All the teeth villiform, in bands; teeth on the palatine bones as well as on the vomer; tongue smooth. The spinous dorsal fin is continuous with the soft, and composed of ten strong spines; three anal spines. Operculum with a flat spine; praeoperculum finely serrated behind, and with small denticulations on the lower limb; praeorbital serrated. Scales small, strongly ctenoid.
Macquarie River   A. Gunther[52]

The drawing that accompanied Gunther's description of 1871.[52] The golden perch is a valuable sporting and food fish, commonly growing to 700 mm long.

Very beautifully coloured   1883
The Golden Perch or Yellow-belly represents two species of *Ctenolates*, *C. ambiguus* and *C. christyi*, a species described by Count Castelnau, from the Edwards River. The first of these is common in all the rivers and lagoons of the interior. It is a very rich and delicate fish, and attains a weight of 7 lbs. or more; its time and manner of spawning is the same as the cod. The spawn is believed to be hatched in a fortnight after deposition. When fresh this fish is coloured very beautifully. The body is of a magnificent green, the sides and behind the dorsal, the upper parts of the body, are rich golden. The head is a beautiful mixture of green, purple, yellow, and scarlet, with fine golden tinges; the belly is white, the dorsal fin purplish green, anal scarlet, with its base yellow and its end purple, pectorals scarlet at their base, and yellow in their second half; the eye is purple, with an interior white ring. These colours are subject to great variation, and the belly is sometimes red. The young fish have little of the fine hues of the adult, and they are much more elongate. The head is purple, and the dorsal fin is grey, bordered with black.
New South Wales   J. E. Tenison-Woods[119]

Sources

The material that formed the basis of this book is located in various public libraries, museums and private collections. The key to the locations is:

AM = The Australian Museum Library.
BM = British Museum (Natural History).
ML = Mitchell Library, State Library of New South Wales.
MM = Macleay Museum, The University of Sydney.
NL = National Library of Australia, Canberra.
PC = private collection.
SU = Fisher Library, The University of Sydney.

ALEXANDER, W. B. 'History of Zoology in Western Australia, Pt II, 1791-1829'. *Journal and Proceedings of the Royal Society of Western Australia*, Vol. 1. 1914-15. (SU)

ALEXANDER, W. B. 'Note on the Birds met with on the Swan River by Vlamingh in 1697'. *Journal and Proceedings of the Royal Society of Western Australia*, Vol. 1. 1914-15. (SU)

1 ANDERSON, W. quoted in BEAGLEHOLE, J. C. *The Journals of Captain James Cook on his Voyages of Discovery*, Cambridge: University Press, for Hakluyt Soc., 1961. (SU)

2 ANON. *The Naturalist's Pocket Magazine*, Vol. II. 1799. (PC)

3 ANON. 'The Worlds Oldest Bird' in POLLARD, J. *Birds of Paradox*, Melbourne: Lansdowne Press, 1967. (SU)

AUSTRALIAN MUSEUM. *A Catalogue of the Specimens of Natural History and Miscellaneous Curiosities deposited in the Australian Museum*. Sydney: Tegg & Co., 1837. (AM)

4 BARRALLIER, F. L. quoted in *Historical Records of New South Wales.*, Vol. V. Appendix. (SU)

5 BARRINGTON, G. *The History of New South Wales including Botany Bay, Port Jackson, Parramatta, Sydney and all its Dependancies From the Original Discovery of the Island, with the Customs and Manners of the Natives and an Account of the English Colony from its Foundation to the Present Time*. London: M. Jones, 1802. (SU)

BARRINGTON, G. *An Account of a Voyage to New South Wales . . . to which is Prefixed a Detail of his Life, Trials, Speeches &c. Enriched with Beautiful Color'd Prints*. London: M. Jones, 1810. (SU)

BARTON, G. B. *History of New South Wales from the Records*. Vol. 1. Sydney: Potter, 1889. (SU)

6 BASS, G. quoted in COLLINS, D. *An Account of the English Colony in New South Wales*. Vol. 2. London: Cadell & Davies, 1802. (SU)

7 BEAGLEHOLE, J. C. *The Journals of Captain James Cook on his Voyages of Discovery*. 3 vols in 4 books and portfolio. Cambridge: University Press, for Hakluyt Soc., 1961. (SU)

8 BEAGLEHOLE, J. C. *The Endeavour Journal of Joseph Banks 1768-1771*. 2 vols. Sydney: Public Library and Angus & Robertson, 1962. (SU)

9 BELL, E. A. *A Historic Centenary – Roberts, Stewart & Co. Ltd., 1865-1965*. Hobart: Mercury Press, 1965. (ML)

10 BENNETT, G. *Wanderings in New South Wales; Batavia, Pedir Coast, Singapore, and China: Being the Journal of a Naturalist in those Countries, During 1832, 1833, and 1834*. 2 vols. London: Richard Bentley, 1834. (SU)

11 BENNETT, G. *Gatherings of a Naturalist in Australasia: Being Observations Principally on the Animal and Vegetable Productions of New South Wales, New Zealand and Some of the Austral Islands*. London: J. van Doorst, 1860. (SU)

12 BERNCASTLE, Dr. 'On the Distinction Between the Harmless and Venomous Snakes of Australia'. *The Australian Medical Gazette*. February 1870. (SU)

13 BEWICK, T. *A General History of Quadrupeds*. 4th edn. Newcastle-upon-Tyne: Bewick, 1805. (SU)

14 BINGLEY, Rev. W. *Animal Biography*. London: R. Phillips, 1805. (PC)

15 BLACKBURN, D. A. A letter to Mr Richard Knight, Devizes, England. *Journal of the Royal Australian Historical Society*. Vol. 20. Sydney, 1934. (SU)

16 BLACK SWAN. Davenport plate, c. 1820-30, with inscription on back 'Black Swan of New South Wales', in SHAW, G. *The Naturalist's Miscellany*.

17 BLAND, W. A letter to the editor. *Sydney Gazette*. 28 October 1826. (ML)

18 BLUNT, W. *The Ark in the Park*. London: Hamilton and the Tryon Gallery, 1976. (PC)

BOUGAINVILLE, L. A. de. *A Voyage around the World Performed by Order of His Most Christian Majesty in the Years 1766, 1767, 1768 and 1769 by Lewis de Bougainville . . . in the Frigate La Boudese and the Storeship L'Etoile*. Translated by J. R. Forster. London: Nourse & Davies, 1772. (SU)

BOULENGER, G. A. *Catalogue of Lizards in the British Museum (Natural History)*. 2nd edn. 3 vols. London: Order of the Trustees, 1885-87. (SU)

BOULENGER, G. A. *Catalogue of Fishes in the British Museum (Natural History)*. London: Order of the Trustees, 1895. (SU)

BOULENGER, G. A. *Catalogue of Snakes in the British Museum (Natural History)*. 3 vols. London: Order of the Trustees, 1893-96. (SU)

BOWDLER SHARPE, R., GADOW, H., SCLATER, P. L., SALVADORI, T. *et al. Catalogue of the Birds in the British Museum*. 27 vols. London: Order of the Trustees, 1881-98. (SU)

19 BOWES, A. Original daily journal kept on the transport *Lady Penrhyn* 1788. MS (ML)

20 BRANDL, E. J. Australian Aboriginal Paintings. *Australian Aboriginal Studies*. No. 52 (Prehistory and Cultural Series, No. 9) Canberra: Australian Institute of Aboriginal Studies, 1973. (PC)

21 BROWN, P. *New Illustrations of Zoology, Containing Fifty Coloured Plates of New, Curious and Non-Descript Birds, with a few Quadrupeds, Reptiles and Insects, together with a Short and Scientific Description of the Same*. London: White, 1776. (AM)

22 BURNEY, J. quoted in BEAGLEHOLE, J. C. *The Journals of Captain James Cook on his Voyages of Discovery*. Cambridge: University Press, for Hakluyt Soc., 1961. (SU)

23 CAEN, A. quoted in HEERES, J. E. *The Part Borne by the Dutch in the Discovery of Australia 1606-1765*. Leiden: Brill, 1899. (SU)

24 CALDWELL, W. H. 'The Embryology of Monotremata and Marsupialia, Pt 1'. *Philosophical Transactions of the Royal Society*. Vol. 179, 1887. (SU)

25 CALEY, G. quoted in VIGORS, N. A. and HORSFIELD, T. A. Description of the Australian Birds in the Collection of the Linnean Society: With an Attempt at Arranging Them According to their Natural Affinities. *Transactions of the Linnean Society of London*. Vol. XV. 1827. (MM)

26 CAMPBELL, W. D. 'Aboriginal Carvings of Port Jackson and Broken Bay'. *Memoirs of the Geological Survey of New South Wales*. Sydney: Department of Mines and Agriculture, 1898. (PC)

27 CARSTENZOON, J. quoted in HEERES, J. E. *The Part Borne by the Dutch in the Discovery of Australia 1606-1765*. Leiden: Brill, 1899. (SU)

28 COLLINS, D. *An Account of the English Colony in New South Wales*. 2 vols. London: Cadell & Davies, 1798-1802. (SU)

29 CROZET, J. *Crozet's Voyage to Tasmania, New Zealand and the Ladrone Islands and the Philippines in the years 1771-1772*. Translated by H. Ling Roth. London: Truslove & Shirley, 1891. (SU)

CUNNINGHAM, P. *Two Years in New South Wales: A Series of Letters*. 2 vols. London: Colburn, 1827. (SU)

30 DAMPIER, W. *Voyages*. 4 vols. London: Knapton, 1703-9. (SU)

31 DAVIES, T. 'Description of *Menura superba*, a Bird of New South Wales'. *Transactions of the Linnean Society of London*. Vol. VI. 1802. (MM)

32 de BRUIN, C. *Reizen over Muskovie door Persie en Indie*. 1714. (PC)

33 de JODE, C. *Speculum Orbis Terrae*. Antwerp, 1593. (ML)

34 de VIS, C. 'Myology of Chlamydosaurus Kingii'. *Proceedings of the Linnean Society of New South Wales*. 8. 1884. (MM)

35 DUMERIL, A. M. C. *Erpetologie Générale ou Histoire Naturelle Complète des Reptiles*. Paris: Librairie Encyclopédique de Roret, 1854. (MM)

36 ELLIS, W. drawing No. 95. 1777. Reproduced by permission of the Trustees of the British Museum (Natural History). (BM)

36a ETHERIDGE, R. *et al. Lord Howe Island: Its Zoology, Geology, and Physical Characters*. Sydney: Government Printer, 1889. (SU)

EYRE, J. E. *Journals of Expeditions of Discovery into Central Australia and Overland from Adelaide to King George's Sound in the Years 1840-1 . . . Including an Account of the Manners and Customs of the Aborigines and the State of their Relations with Europeans*. London: T. & W. Boone, 1845. (SU)

FLOWER, W. H. and GARSON, J. G. *Catalogue of the Specimens Illustrating the Osteology and Dentition of Vertebrated Animals, Recent and Extinct, Contained in the Museum of the Royal College of Surgeons of England*. Pt II. London: printed for the College, 1884. (SU)

37 © FORSHAW, J. M. and COOPER, W. T. *The Birds of Paradise and Bowerbirds*. Sydney: Collins, 1977. Reproduction of bower sketch by kind permission of the artist and the publishers William Collins Pty Ltd. (SU)

FORSTER, J. G. A. *A Voyage Round the World in His Britannic Majesty's Sloop Resolution, commanded by Capt. James Cook, During the Years 1772, 3, 4, and 5 by George Forster*. London: White, 1777. (SU)

38 FOX, G. T. *Synopsis of the Newcastle Museum, Late the Allan, Formerly the Tunstall, or Wycliffe Museum; To which are Prefixed Memoirs of Mr. Tunstall . . . and of Mr. Allan . . . with Occasional Remarks on the Species . . .* Newcastle: T. & J. Hodgson, 1827. (ML)

FRITH, H. J. and CALABY, J. H. *Kangaroos*. Melbourne: Cheshire, 1969. (SU)

39 GILBERT, J. Manuscript of his diary kept on the Leichhardt expedition of 1844-45. MS. (ML)

40 GILBERT, T. *Voyage from New South Wales to Canton, in the Year 1788, With Views of the Islands Discovered*. London: Stafford, for Debrett, 1789. (SU)

GMELIN, J. F. *Systema Naturae*. 10 vols. 13th edn. Lipsiae: Beer, 1788-93. (AM)

41 GOLDSMITH, Oliver. *History of the Earth and Animated Nature*. Glasgow: Blackie & Son, n.d. (1840?). (PC)

GOULD, J. Letters and papers of John Gould and John Gilbert, 1839, 1842-43. MSS. (ML)

42 GOULD, J. *The Birds of Australia*. 36 parts in 7 vols. London: the author, 1840-48. (SU)

GOULD, J. *The Birds of Australia*. Supplement. 5 parts in 1 vol. London: the author, 1851-69. (SU)

43 GOULD, J. 'On a New Species of Platycercus'. *Annals and Magazine of Natural History*. 15, 1845. (SU)

44 GOULD, J. *The Mammals of Australia*. 3 vols. London: the author, 1845-63. (SU)

45 GOVETT, W. R. 'Sketches of New South Wales'. *The Sunday Magazine*. 1836. (ML)

46 GRAY, J. E. quoted in KING, P. P. *Narrative of a Survey of the Intertropical and Western Coasts of Australia Between the Years 1818 and 1822* . . . 2 vols. London: Murray, 1827. (SU)

47 GRAY, J. E. quoted in GREY, G. *Journals of Two Expeditions of Discovery in North-west and Western Australia During the Years 1837, 38 and 39* . . . London: T. & W. Boone, 1841. (SU)

48 GREY, G. *Journals of Two Expeditions of Discovery in North-west and Western Australia During the Years 1837, 38 and 39* . . . *Describing Many Newly Discovered Important and Fertile Districts, With Observations on the Moral and Physical Condition of the Aboriginal Inhabitants, &c. &c.* London: T. & W. Boone, 1841. (SU)

49 GRIGG, G. C. 'Spawning Behaviour in the Queensland Lungfish, *Neoceratodus forsteri*'. *Australian Natural History*, 15. No. 3. Sydney: Australian Museum, 1965. (PC)

50 GUNTHER, A. *Catalogue of Fishes in the British Museum (Natural History)*. Vol. VIII. London: Trustees of the British Museum, 1870. (SU)

51 GUNTHER, A. 'Description of *Ceratodus*, a Genus of Ganoid Fishes, Recently Discovered in the Rivers of Queensland, Australia'. *Philosophical Transactions of the Royal Society of London*. 1871. (SU)

52 GUNTHER, A. 'Description of a New Percoid Fish from the Macquarie River'. *Proceedings of the Zoological Society of London*. 1871. (SU)

53 GUNTHER, A. *Catalogue of the Lizards in the British Museum (Natural History)*. Vol. 1. London: Trustees of the British Museum, 1885. (SU)

54 GUNTHER, A. 'Australian Reptiles, 1789-1897' in a scrapbook, Illustrations Relating to Australian Fauna. MS. (ML)

55 HARRIS, G. P. 'Description of Two New Species of Didelphis from Van Diemen's Land'. *Transactions of the Linnean Society of London*. Vol. IX. 1807. (MM)

56 HEERES, J. E. *The Part Borne by the Dutch in the Discovery of Australia, 1606-1765*. Leiden: Brill, 1899. (SU)

57 HOME, E. 'A Description of the Anatomy of the *Ornithorhynchus hystrix*'. *Philosophical Transactions of the Royal Society of London*. 1802. (SU)

HOME, E. 'A Description of the Anatomy of the *Ornithorhynchus paradoxus*'. *Philosophical Transactions of the Royal Society of London*. 1802. (SU)

58 HOME, E. 'An Account of Some Peculiarities in the Anatomical Structure of the Wombat, with Observations on the Female Organs of Generation'. *Philosophical Transactions of the Royal Society of London*. 1808. (SU)

59 HUNTER, J. *An Historical Journal of the Transactions at Port Jackson and Norfolk Island, With the Discoveries which have been made in New South Wales and in the Southern Ocean Since the Publication of Phillip's Voyage* . . . London: Stockdale, 1793. (SU)

60 HUNTER, J. Letter to Sir Joseph Banks. quoted in WHITLEY, G. P. *More Early History of Australian Zoology*. Sydney: Royal Zoological Society of New South Wales, 1975. (SU)

61 HUNTER, J. Sketchbook in the Rex Nan Kivell Collection, National Library of Australia. (NL)

JOUANIN, C. 'Josephine and the Natural Sciences'. *Apollo*. July 1977. (PC)

62 KING, J. *A Voyage to the Pacific Ocean . . . Performed under the Direction of Captains Cook, Clerke and Gore in His Majesty's Ships the Resolution and Discovery in the Years 1776, 1777, 1778, 1779, 1780*. London: Strahan, for Nicol & Cadell, 1784. (SU)

63 KING, P. P. *Narrative of a Survey of the Intertropical and Western Coasts of Australia Between the Years 1818 and 1822 . . . With an Appendix containing Various Subjects Relating to Hydrography and Natural History*. 2 vols. London: Murray, 1827. (SU)

64 KNIGHT, C. *The Pictorial Museum of Animated Nature*. London, Knight & Co., n.d. (1860?). (PC)

65 KREFFT, G. *The Snakes of Australia: An Illustrated and Descriptive Catalogue of all the Known Species*. Sydney: Government Printer, 1869. (SU)

66 KREFFT, G. Letter to the editor. *Sydney Morning Herald*. 17 January 1870. (ML)

67 KREFFT, G. 'Description of a Gigantic Amphibian Allied to Genus *Lepidosiren* From the Wide Bay District, Queensland'. *Proceedings of the Scientific Meetings of the Zoological Society of London*. 2. 1870. (SU)

KREFFT, G. *The Mammals of Australia*. Sydney: Government Printer, 1871. (SU)

68 LA BILLARDIERE, J. J. H. de. *An Account of a Voyage in Search of La Pérouse Undertaken by Order of the Constituent Assembly of France, and Performed in the Years 1791, 1792, and 1793 . . . Under the Command of d'Entrecastenaux*. London: Debrett, 1800. (SU)

LAKE, J. S. *Freshwater Fishes and Rivers of Australia*. Melbourne: Nelson, 1971. (SU)

LATHAM, J. *A General Synopsis of Birds*. 3 vols. London: White, 1781-85. (AM)

69 LATHAM, J. *Supplement to A General Synopsis of Birds*. London: Leigh & Sotheby, 1787. (AM)

LATHAM, J. *Index Ornithologicus* . . . 2 vols. London: for the author, 1790. (AM)

70 LEE, I. *Captain Bligh's Second Voyage to the South Sea*. 1920. (SU)

71 LEICHHARDT, F. W. L. *Journal of an Overland Expedition in Australia from Moreton Bay to Port Essington, a Distance of Upwards of 3000 Miles During the Years 1844-1845*. London: T. & W. Boone, 1847. (SU)

72 LESSON, R. P. *Voyage Autour du Monde Enterpris par Ordre De Gouvernement sur la Corvette La Coquille*. 2 vols. Paris: Pourrat Frères. (AM)

LEWIN, J. W. *Birds of New Holland With their Natural History, Collected, Engraved and Faithfully Painted After Nature*. London: White & Bagster, 1808. (NL)

73 LEWIN, J. W. *A Natural History of the Birds of New South Wales, Collected, Engraved and Faithfully Painted After Nature*. London: Bohte, 1822. (SU)

74 LISTER, M. Part of a Letter from Mister Witsen, Burger Master of Amsterdam, and F.R.S., to Dr Martin Lister, fellow of the College of Physicians, and R.S., Concerning some later Observations in Nova Hollandia. *Philosophical Transactions of the Royal Society of London*. 1698. (SU)

75 LUMHOLTZ, C. *Among Cannibals: An Account of Four Years' Travels in Australia and of Camp Life with the Aborigines of Queensland*. London: Murray, 1889. (SU)

76 Lungfish – 'A Haul of Lungfish' – original historical photograph. (AM)

77 MacGILLIVRAY, J. *Narrative of the Voyage of the H.M.S. Rattlesnake Commanded by the Late Captain Owen Stanley . . . during the Years 1846-1850, Including Discoveries and Surveys in New Guinea, the Louisiade Archipelago etc., to which is added the Account of Mr. E. B. Kennedy's Exploration of the Cape York Peninsula*. 2 vols. London: T. & W. Boone, 1852. (SU)

78 McKAY, J. (ed.) *The Rosella Story 1895-1963*. Melbourne: The Rosella Preserving Co. Ltd, 1977. (MM)

79 MAHONY, D. J. 'On the Bones of the Tasmanian Devil and Other Animals, Associated with Human Remains near Warnambool: With a Note on the Dune Sand'. *Victorian Naturalist*, Vol. XXIX. July 1912. (ML)

MAJOR, R. (ed.). *Early Voyages to Terra Australis, now called Australia: A Collection of Documents and Extracts from Early Manuscript Maps, Illustrative of the History of Discovery on the Coasts of That Vast Island, From the Beginning of the Sixteenth Century to the Time of Captain Cook*. London: Hakluyt Soc., 1859. (SU)

80 MARETT, F. J. 'Fishing for Murray Cod' in *Walkabout*. 1 December 1943. (ML)

MARTIN, C. J. 'Venom of the Australian Blacksnake

(*Pseudechis porphyriacus*)'. *Journal of the Royal Society of New South Wales*. 1892. (ML)

MASEFIELD, J. (ed.). *Dampier's Voyages, by William Dampier*. 2 vols. London: Grant Richards, 1906. (SU)

81 MATHEWS, G. M. *The Birds of Norfolk and Lord Howe Islands and the Australasian South Polar Quadrant, with additions to 'The Birds of Australia'*. London: H. & G. Witherby, 1928. (SU)

MATHEWS, G. M. *A List of the Birds of Australasia (including New Zealand, Lord Howe and Norfolk Islands and the Australasian Antarctic Quadrant)*. London: for the author, by Taylor & Francis, 1931. (SU)

82 MATHEWS, R. H. 'The Aboriginal Rock Pictures of Australia'. *Proceedings and Transactions of the Queensland Branch of the Royal Geographical Society of Australia*. 1895. (SU)

MILET-MUREAU, L. M. A. D. *Voyage de la Pérouse autour de monde*. Paris: de l'imprimerie de la République, 1797. (SU)

83 MITCHELL, T. L. *Three Expeditions into the Interior of Eastern Australia: With Descriptions of the Recently Explored Region of Australia Felix, and of the Present Colony of New South Wales*. 2 vols. 2nd edn. London: T. & W. Boone, 1839. (SU)

MOUNTFORD, C. P. *The Dreamtime Book: Australian Aboriginal Myth in Paintings by Ainslie Roberts*. Adelaide: Rigby, 1973. (PC)

New London Magazine, The. September 1790. (ML)

PARKER, C. S. *Australian Legendary Tales . . .* London: Nutt; and Melbourne: Mullen & Slade, 1897. (SU)

84 PARKINSON, S. A. *Journal of a Voyage to the South Seas, in His Majesty's Ship, the Endeavour, Faithfully Transcribed from the Papers of the Late Sydney Parkinson, Draughtsman to Joseph Banks, Esq., on his Late Expedition with Dr. Solander Round the World*. London: Parkinson, 1773. (AM)

85 PARKINSON, S.A. Sketch of a female Banksian Cockatoo. 1770. Reproduced by kind permission of the Trustees of the British Museum (Natural History).

86 PENNANT, T. *Genera of Birds*. 2nd edn. London: White, 1781. (AM)

87 *Penny Cyclopaedia of the Society for the Diffusion of Useful Knowledge*. Vol. 17. London: Knight & Co., 1840. (PC)

88 PERON, F. and FREYCINET, L. *Historique Atlas, Voyage de Découvertes aux Terres Australis*. Paris: Imprimerie Imperiale, 1807. (ML)

89 PERRY, G. *Arcana: Or, The Museum of Natural History, Containing the Most Recent Discovered Objects Embellished with Coloured Plates and Corresponding Descriptions, With Extracts Relating to Animals and Remarks of Celebrated Travellers, Combining a General Survey of Nature.* London: Stafford, 1810-11. (NL)

90 PHILLIP, A. *The Voyage of Governor Phillip to Botany Bay, With an Account of the Establishment of the Colonies of Port Jackson & Norfolk Island Compiled from Authentic Papers . . . To which are added the Journals of Lieuts. Shortland, Watts, Ball & Capt. Marshall, With an Account of their Discoveries.* London: Stockdale, 1789. (SU)

POIGNANT, A. *The Improbable Kangaroo and Other Australian Animals.* Sydney: Angus & Robertson, 1965.

91 PORTLOCK, N. quoted in LEE. I. *Captain Bligh's Second Voyage to the South Sea.* 1920. (SU)

92 PRICE, J. quoted in *Historical Records of New South Wales.* 3. Appendix C. (SU)

QUOY, J. R. C. and GAIMARD, J. P. *Voyage de 'L' Astrolabe Pendant les Anées 1826-27-28-29 sous le Commandement de M. Dumont D'Urville.* Paris, 1830. (SU)

RAWSON, G. *Matthew Flinder's Narrative of his Voyage in the Schooner Francis 1798 . . .* London: Golden Cockerel Press, 1946. (SU)

93 ROPER, E. 'Kangaroo Hunt, Mount Zero, The Grampians, Victoria' 1880 (oil) in Rex Nan Kivell Collection, NK 1169, National Library of Australia. (NL). Reproduction courtesy of the Trustees, National Library of Australia.

94 ROTH, W. E. 'North Queensland Ethnology'. *Records of the Australian Museum.* Vol. VII. 1908. (AM)

95 SCLATER, P. L. 'A Report to the Zoological Society of London'. *Proceedings of the Zoological Society of London,* Vol. XIV. 1886. (SU)

96 SCOTT, T. Sketch of 'Tyger Trap'. Original watercolour in Mitchell Library. (ML). Reproduction courtesy of the Trustees of the Mitchell Library.

97 SHAW, G. *The Naturalist's Miscellany: Or, Coloured Figures of Natural Objects Drawn & Described Immediately from Nature.* 24 vols. London: Nodder, 1789-1813. (AM)

98 SHAW, G. *Zoology and Botany of New Holland.* London: Sowerby, 1793. (ML)

99 SHAW, G. 'Description of a new Species of Mycteria'. *Transactions of the Linnean Society of London.* Vol. V. 1798. (MM)

100 SHAW, G. *General Zoology; Or Systematic Natural History.* 14 vols. London: Kearsley, etc., 1800-26. (AM)

SHERBORN, C. D. 'On the Dates of Shaw & Nodder's "Naturalist's Miscellany" '. *Annals and Magazine of Natural History.* Series 6, Vol. XV. 1895. (SU)

101 SONNERAT, P. *Voyage à la Nouvelle Guinée, dans lequel on trouve la description des lieux, des observations physiques & morales, & des details relatifs à l'histoire naturelle dans le règne animal & le règne végètal.* Paris: Ruault, 1776. (ML)

STANBURY, P. J. *Looking at Mammals.* Sydney: Angus & Robertson, 1970. (MM)

STANBURY, P. J. (ed.). *One Hundred Years of Australian Scientific Explorations.* Sydney: Holt-Saunders, 1975. (MM)

102 STEIGLITZ, K. R. von (ed.). *Sketches in Early Van Diemen's Land by Thomas Scott, Assistant Surveyor General of Van Diemen's Land 1821-1836.* Hobart: Fullers Book Shop, 1966. (ML)

103 STOKES, J. L. *Discoveries in Australia; With an Account of the Coasts and Rivers Explored and Surveyed During the Voyage of H.M.S. Beagle In the Years 1837-38-39-40-41-42-43 . . .* 2 vols. London: T. & W. Boone, 1846. (SU)

STRZELECKI, P. E. de. *Physical Description of New South Wales and Van Diemen's Land.* London: Longmans, 1845. (SU)

STURT, C. *Two Expeditions into the Interior of Southern Australia During the Years 1828, 1829, 1830 and 1831; With Observations on the Soil, Climate and General Resources of the Colony of New South Wales.* 2nd edn. London: Smith, Elder, 1834. (SU)

104 STURT, C. *Narrative of an Expedition into Central Australia . . . During the Years 1844, 5 and 6; Together with a Notice of the Province of South Australia in 1847.* 2 vols. London: T. & W. Boone, 1849. (SU)

105 *Sydney Gazette.* 21 August 1803 and 9 October 1803. (ML)

105a *Sydney Gazette.* Vol. III, 21 April 1805. (ML)

106 *Sydney Gazette.* 'New Parrot Species?'. 7 August 1804. (ML)

107 *Sydney Gazette.* 'Buzzing Possum'. 5 February 1809. (ML)

108 *Sydney Gazette.* 'Peach Field Affected by Parrots'. 7 April 1821. (ML)

109 TEMMINCK, C. J. *Nouveau Recueil de Planches Coloriées d'Oiseaux Pour Servir de Suite et de Complément aux*

Planches Enluminées de Buffon . . . 1770. Paris: Levrault, 1820-39. (PC)

TENCH, W. *A Narrative of the Expedition to Botany Bay; With an Account of New South Wales, its Productions, Inhabitants, &c., To which is Subjoined a List of the Civil and Military Establishments at Port Jackson.* London: Debrett, 1789. (SU)

110 TENCH, W. *A Complete Account of the Settlement at Port Jackson in New South Wales, Including an Accurate Description of the Situation of the Colony, of its Natives and of its Natural Productions, Taken on the Spot by Captain Watkin Tench of the Marines.* London: Nichol & Sewell, 1793. (SU)

111 TOBIN, G. quoted in WHITLEY, G. P. 'George Tobin, a neglected naturalist'. *Australian Museum Magazine*, 5, No. 2. 1932. (SU)

112 VALENTIJN, F. *Oud En Nieun Oost-Indien.* 5 vols. Amsterdam: Braam Linden, 1724-26. (ML)

113 WATERHOUSE, G. R. *Mammalia, Vol. XI: The Natural History of Marsupialia or Pouched Animals.* Edinburgh: Lizars, 1841; in JARDINE, Sir. W. *The Naturalists Library.* 30 vols. (SU)

114 WATERHOUSE, G. R. *A Natural History of the Mammalia.* 2 vols. London: Bailliere, 1846-48. (SU)

Western Mail. Christmas supplement. 24 December 1897. (ML)

115 WHITE, J. *Journal of a Voyage to New South Wales; with Sixty-five Plates of Non descript Animals, Birds, Lizards, Serpents, Curious Cones of Trees and Other Natural Productions.* London: Debrett, 1790. (SU)

116 WHITLEY, G. P. *An Early History of Australian Zoology.* Sydney: Royal Zoological Society of New South Wales, 1970. (SU)

WHITLEY, G. P. *More Early History of Australian Zoology.* Sydney: Royal Zoological Society of New South Wales, 1975. (SU)

WHITTELL, H. M. *The Literature of Australian Birds.* Perth, W.A.: Paterson Brokenshaw, 1954. (SU)

117 WILLIAMS, R. *Extract of a Journal from England to Botany Bay, by Richard Williams Belonging to the Ship Borrow Dale, Captain Reed.* Handbill posted in back of Mitchell Library copy. (ML)

118 WILSON, C. A. 'Notes on *Moloch horridus*'. *Journal of the Linnean Society Proceedings: Zoology.* Vol. X. 1870. (MM)

119 WOODS, J. E. T. *Fish and Fisheries of New South Wales.* Sydney: Government Printer, 1883. (AM)

Index

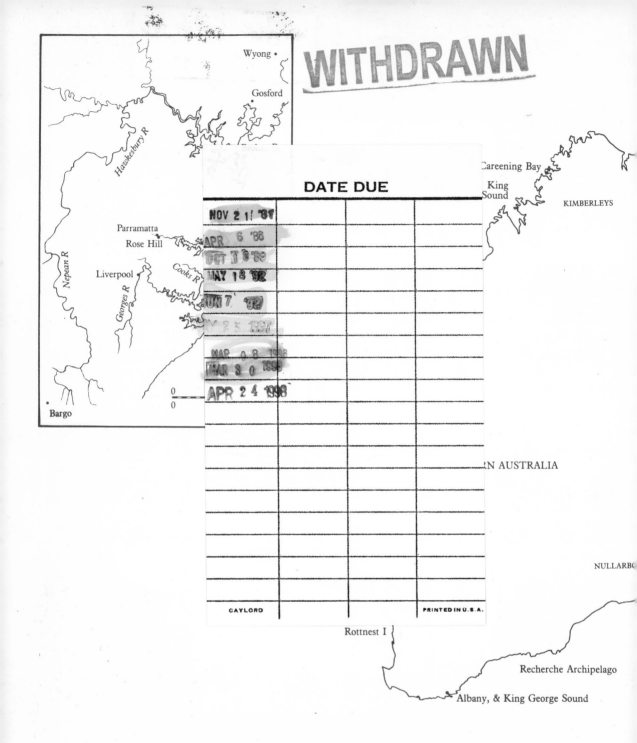

Wyong •

Gosford

Haukesbury R

Careening Bay

King
Sound

KIMBERLEYS

Parramatta
Rose Hill

Nepean R

Liverpool •

Cooks R

Georges R

0
0

Bargo •

RN AUSTRALIA

NULLARB

Rottnest I

Recherche Archipelago

Albany, & King George Sound

0   100 200 300 km

0   100   200 miles